Environmental Planning and Management

# Environmental Planning and Management

## Christian N Madu
*Pace University, New York*

Imperial College Press

*Published by*

Imperial College Press
57 Shelton Street
Covent Garden
London WC2H 9HE

*Distributed by*

World Scientific Publishing Co. Pte. Ltd.
5 Toh Tuck Link, Singapore 596224
*USA office:* 27 Warren Street, Suite 401-402, Hackensack, NJ 07601
*UK office:* 57 Shelton Street, Covent Garden, London WC2H 9HE

**British Library Cataloguing-in-Publication Data**
A catalogue record for this book is available from the British Library.

**ENVIRONMENTAL PLANNING AND MANAGEMENT**

ISBN-13 978-1-86094-671-4
ISBN-10 1-86094-671-2

Editor: Tjan Kwang Wei

Printed in Singapore by World Scientific Printers (S) Pte Ltd

# Preface

The focus of this book is on environmental planning and management. We view planning as a critical component of pollution prevention and also a necessary component of resource optimization. Management of natural resources requires the evaluation of alternative options and the optimal selection of sustainable choices.

As corporations today compete for customers and market share, they are rapidly introducing new products, consuming limited natural resources, creating more waste and pollution. It is important to introduce corporations and their top management to sustainable practices. They become environmentally conscious when they understand the risks associated with poor environmental records. Such risks could be assessed and quantified as costs to the corporation. This book uses the four types of costs: internal, prevention, appraisal and external that are prominent in quality management literature to expose top management to the risks of poor environmental quality. More especially, the external failure cost which includes loss of customer goodwill is usually underestimated but has the greatest potential of affecting the survival of the firm.

Top management must be proactive and take a cradle-to-grave approach for its products. This requires proper life cycle assessment that would document all the inventories associated with the product as well as the products' interactions with the different environmental media: land, air and water.

Competing in today's marketplace requires a holistic view of both products and processes. It requires that companies pay attention to their stakeholders in addition to their customers. Environmental planning lays the foundation to adapt to the needs of the changing world and avoid the hazards, risks and high costs associated with poor environmental practices.

This book identifies good environmental practices, lays down effective strategies and practical models. It focuses on designing for the environment. This is based on the notion that by adopting preventive methods, external failure costs can be minimized. Contrary to the popular belief that the costs of implementing environmental practices are high, we have presented cases of companies that have increased profitability because of their environmental programs. We have presented such practices in Environmental Action Boxes that appear at the end of some chapters. We have also referred to companies that have suffered huge costs and loss of customer goodwill due to their poor environmental practices. This book focuses mostly on using sustainable practices to achieve competitiveness. It refers to companies that have achieved such feat to encourage others to adopt proactive environmental policies. It also points to worldwide focus on international standards especially as outlined in ISO 14000 Environmental Management Systems.

A few of the chapters in this book were revised from my earlier research on environmental management. However, several new chapters on Designing for the Environment, Risk Assessment and ISO 14000, have been included.

This book is intended as an introduction to corporate environmental management and is suitable for basic courses in the subject. Practitioners would also find it helpful as it explains some of the basic concepts and environmental strategies that are in practice today.

Finally, I thank my graduate assistants Kaushal Patel and Nishant Sheth who worked very hard to organize the final manuscript.

**Christian N. Madu**
Lubin School of Business
Pace University, New York

# Contents

Preface                                                                          v

Chapter 1    Introduction to Environmental Planning and                          1
             Management

Chapter 2    Sustainable Manufacturing                                          14

Chapter 3    Environmentally Conscious Manufacturing                            42

Chapter 4    The ISO 14000 Model                                                62

Chapter 5    Environmental Planning                                             90

Chapter 6    Life Cycle Assessment                                             113

Chapter 7    Design for the Environment – Part I                               146

Chapter 8    Design for the Environment – Part II                              163

Chapter 9    Manufacturing Strategies: Agile, Lean and                         190
             Flow Manufacturing

Chapter 10   Environmental Risk Assessment and Management                      209

Chapter 11   Competing on Environmental Management                             228

Index                                                                          241

# Chapter 1

# Introduction to Environmental Planning and Management

Environmental planning and management are strategic challenges confronting businesses in the 21st century. As customers and stakeholders demand that global companies lead by showing corporate citizenship, the health, safety and security of both the people and the natural environment have become a litmus test for good citizenry. Successful companies today are not only measured by how much profitable their products are but also how well they respond to the call to protect the natural environment. Global companies are increasingly recognizing that green products do not necessarily cost more but provide market niche that could help the company thrive. For example, the success stories of Kodak single-use camera and Xerox re-manufacturing programs have made businesses rethink their strategies and pay attention to environmental management.

The issues concerning the environment are not purely scientific but rely also on planning and management. They are part of corporate vision, mission, and strategy which need to be effectively planned to remain competitive. Corporate leaders are recognizing the increasing role of stakeholders and are accepting that their corporate strategies must focus not only on their customers but also on their stakeholders. Stakeholders are those whose actions or reactions have the potential to affect the firm's business operation and survival. Environmental burden is an issue that affects world citizens and they have the right to demand that corporate responsibility be properly defined to take into consideration these concerns. Today, companies are adapting their strategies in response to these concerns. World communities have reacted by adopting international and national laws to ensure compliance to environmental standards. It is also clear that green products may in fact, provide competitive advantage. Recently, there has been significant growth in the development of such products since the 1990s.

Paint companies are now producing zero-VOC (Volatile Organic Compounds) products. Glidden introduced zero-VOC paint in the US in the mid 1990s and was followed by other major paint manufacturers such as Benjamin Moore and Sherman-Williams. Homeowners have responded very well to this development[1]. Today, all major paint manufacturers market zero-VOC paint and offer lower-VOC coatings for a variety of substrates. Some manufacturers such as Rodda Paint of Portland, OR; have gained Green Seal on their paint products and that has given it a competitive edge. It joins the likes of big players like Benjamin Moore.

The energy industry has also reacted to *green power* which has been defined by the National Association of Attorney Generals [NAAG] as the use of replenishable or sustainable fuel sources in the generation and transmission of electricity and the disposal of spent fuels. These releases into the environment would not create harmful substances and would pose no significant concern to the ecosystem and to land use. The focus is on renewable energy supply. Studies have shown that homeowners are willing to pay more to use green power[2]. This program is similar to the waste recycling program in many communities that are paid for by consumers.

The trend toward green is on the rise in every sector of the economy. According to a study by Yanklovich Clancy Schulman, 78% of people are "influenced greatly" to buy products that make environmental claims. The Organic Trading Association notes that products with the word "organic" have 34% sell-through rate compared to conventional products. Furthermore, the rate of growth of natural or organic food has been steady at 18-25% while conventional foods remain flat at 3-4%[3]. The demand for organic products has also affected the clothing industry where the current trend calls for the use of organic cotton.

---

[1]  Esposito, C.C., "The evolution of low- and zero-VOC paint," http://www.coatingsworld.com/May041.htm

[2]  PriceWaterHouseCoopers, "The Pitfalls and Potential of Marketing Green Power," Fall 1999 edition of Public Utility Topics, http://www.pwcglobal.com/extweb/indissue.nsf/DocID/45CDA2C5CCA82EE9 8525689B00732FC5

[3]  Fassa, L., "The Trend Toward Green," http://www.babyshopmagazine.com/fall02/green.htm

The growth of *"conspicuous conservation"* is also on the rise as the demand for hybrid vehicles skyrocket[4]. Companies like Toyota and Honda that were among the first to manufacture hybrid cars that combine gas engines with battery-powered electric motors have witnessed surge in demand. These products are not only environmentally friendly but are also economical. Aside from the fact that these hybrid cars cut emissions, owners in the US can write off a one-time deduction of $2,000 of the purchase price for these vehicles since the Internal Revenue Services (IRS) has recognized them as eligible for the clean-burning fuel tax deduction[5]. This law also applies to other vehicles that operate on natural gas, liquefied natural gas, liquefied petroleum gas, hydrogen, or any other fuel that is at least 85% alcohol. Owners of completely electric vehicles would even get tax credit of up to $4,000. Customers who patronize these products want to be seen as saving energy and believe that their consumption behavior reflect who they are. The trend towards environmental conservation is further reflected in the emergence of energy saving products such as energy star appliances, compact fluorescent lights, photovoltaic solar cells, "high performance" homes and others.

The green trend is on the rise and green products offer a new market niche to global companies. According to The LOHAS journal, about 63 million consumers or about 30% of US adults purchase goods and services that are classified as healthy, environmentally friendly, socially conscious and sustainable[6]. The LOHAS index constitutes of five major areas that comprise sustainable economy, healthy living, alternative health care, personal development and ecological life style. While there are more to this index, however, environmental issues are prominent.

---

[4]  Stafford, E.R., "Conspicuous Conservation – Believing that "you are what you own," an emerging trend exalts virtue over tawdry materialism," http://www.greenatworktoday.com.
[5]  Bell, K., "Cut emissions and your tax bill," http://biz.yahoo.com/brn/050309/9775_1.html, March 9, 2005.
[6]  http://www.lohasjournal.com/nbp/app/cda/nbp_cda.php?command=Page& pageType=About

The discussions above highlight the growing importance of environmental management systems and the need to effectively plan and manage in order to harness the values that would accrue to companies that adopt environmental strategies. We shall next look at how to assess the cost of environmental quality. It is very important that managers understand the cost side of environmental quality. Knowledge of the costs would obviously, hit home the need to be proactive and prevent environmental errors from occurring.

## Environmental Management Systems (EMS)

The focus of this book is on Environmental Management and Planning. Naturally, Environmental Management System (EMS) forms the core of any environmental management and planning program. EMS has become increasingly popular in the past few years mainly due to ISO 14000 series which is labeled EMS. EMS is a well structured and documented approach to respond to environmental challenges by focusing on environmental regulations and standards, and customer requirements. It is based on the need to respond proactively to the management of natural environment by conducting lifecycle assessment to lessen environmental burden and optimize the use of limited natural resources. Firms that adopt the EMS approach focus on effective ways to use such limited resources to produce environmentally conscious products and services. One of the popular approaches to achieve this is to adopt a plan-do-check (study) - act (PDCA) which was popularized by the father of Total Quality Management (TQM) Dr. Edward Deming. Using PDCA requires that the firm focuses on how to develop environmental policies by recognizing corporate, industry, local, national and international environmental policies and standards. The consideration of these policies and standards constitutes a major factor in designing a product or service for the environment.

Planning is important to any successful EMS program. A chapter in this book is devoted to the importance of planning. Effective planning leads to successful consideration of factors that may support or hamper

EMS efforts. Planning offers a way to optimization of resources and stops the firm from *"fighting fires."* Planning leads to preventive measures and anticipation of future environmental changes. It positions the firm to take competitive advantage rather than responding only to challenges as they occur.

Effective planning also smoothens out implementation and minimizes environmental burden since many of the sources of environmental burden would be anticipated and addressed. It also helps address *"what if"* or sensitivity analysis questions. Thus, different scenarios and alternatives will be anticipated, investigated, and prototypes looked at to identify at least a *satisficing* solution that may not necessarily be optimal but give a very good resolution to the environmental problem. When products and services are designed for the environment, they are properly planned and fully tested before introduction into the marketplace. However, pushing these products and services to the market does not end the responsibility of the firm. Environmental scanning is an ongoing process that would lead to continuous review of the products and services to achieve continuous improvement.

Our approach in this book is to focus on management issues as key to the success of environmental planning program. The traditional focus of environmental systems has been on the natural sciences. While natural sciences continue to play a major role in understanding the natural environment, the problem of environmental burden or optimization of natural resources would not be achieved without a focus on how such burden or resources can be effectively managed.

This book therefore focuses on the strategic context of environmental management. It focuses on corporate strategies to develop a new vision of environmental management systems. Some of the areas that are looked into are:

- Top management commitment – Since the era of TQM, it has become apparent that most problems organizations face can be blamed on top management. Top management wields a lot of power and authority. It provides the mission and vision of the organization. It lays down organizational strategies and it provides the resources to achieve organizational goals. If top management is not committed to a cause, it would not devote the needed resources to it. Top management must take EMS to heart for it to be successfully implemented. Employees respond to the call from their leaders. Corporate leaders have the responsibility to educate their employees and make them buy into the corporate mission. When top management views environmental management as strategic, integrate it in corporate mission and devote resources and commit time to it, others will follow. These resources would enable to provide proper training, education, and technology to address environmental burden.

- Cross-functional team – There is a need to develop cross-functional teams that would work concurrently to address environmental issues. The use of cross-functional teams is vogue in managing successful organization. It is important that members of the teams have a single view of the organization. Functional units within organizations are interdependent and should not be treated as independent silos. Members of the manufacturing unit should be able to work together with marketing and express and share views at the same time. The concurrency at the design stage helps to reduce the cycle time to introduce new products and makes the organization more competitive. The different backgrounds of team members also expose different worldviews that could be captured and used to develop an acceptable plan to all the departments.

- Stakeholder Teams – In conjunction with the cross-functional teams that may comprise mostly of organizational employees, it is important to also work with stakeholder teams. Stakeholders constitute those whose actions and reactions affect or are affected by organizational actions. As cited in the book, many organizations are today working with various interest groups to develop environmental strategies. The involvement of different stakeholder teams also helps win acceptance of final decisions that are reached.

- Responsibility for the environment: – Who is responsible for environmental management in the organization? Our response to this question is that everyone is responsible. The approach adopted in this book is known as Total Environmental Quality Management (TEQM). Environmental management is not relegated to a particular department or to particular individuals in the organization. Instead, every member of the organization takes full responsibility for his or her own action and works hard to make sure that EMS is fully implemented. Each member of the organization seeks out ways to contribute, and identifies ways to limit pollution and minimize waste in whatever function or activity he or she is involved in.

- The bottom line – Businesses operate to maximize stockholders wealth and must pay attention to the bottom line. If an activity is not value-adding, then there is no need to continue with such an activity. In the past, many corporate leaders complained about the *high cost* of environmental management to justify inaction or disinterest. Today, things have changed. Businesses are observing that the *external cost* for environmental quality has been underestimated and that such costs significantly affect the survival of the business. In addition, they are acknowledging that being environmentally conscious is being competitive. In this book, we illustrate several case studies of companies who regained market shares because they developed successful environmental management programs. The recognition of the importance of environmental management as a competitive weapon could also explain the growing interest of companies to get environmental certifications such as ISO 14000 and the other environmental seals. Companies flout such recognition in advertisement campaigns because they understand that social consciousness influences purchasing decisions. Therefore, it pays to be green.

## The Road to EMS

While there is a need to develop EMS, this road is not always smooth sailing. Companies are often bewildered by the avalanche of materials and guidelines on environmental laws. There is a need for universal standards that would overcome some of the national laws. These standards need to be simplified and designed with implementation in mind. ISO focuses on such issues. However, its guidelines still need to be trimmed to make it easier to implement.

Unfortunately many countries such as the United States are too litigious. This poses a risk to companies who would actually want to identify and rectify problems with their products and processes. There is need to protect such corporate activities as environmental auditing. With the free access to information coupled with "cradle-to-grave" responsibilities for organizational products and services, environmental auditing that detects serious environmental flaws may actually present serious problems to a firm. The government may need to wade in to encourage and protect firms that do environmental auditing to improve environmental quality.

Environmental management should be required for all businesses irrespective of size. Small and medium-sized firms may have more problems due to limited resources and may also lack the knowledge to effectively manage the environment. Government initiatives and incentives may be necessary to encourage such businesses.

Adoption of EMS does not replace sound business strategies. While it would complement the efforts, however, sound business strategies that focus on the organization's core competencies would still be needed to survive in today's highly competitive environment. Thus, the goal should be to include environmental management system as part of the strategic framework. Such consideration would help to ensure that

- Products and services are designed for the environment. In other words, such products and services will create minimal environmental burdens, optimize limited natural resources, and satisfy the growing needs of stakeholders for sustainable or environmentally conscious products and services.
- The firm is competitive and able to survive in its market environment. As mentioned above, auto manufacturers that produced hybrid vehicles have reaped tremendous profits while

producing green products; Kodak repositioned itself with the single-use camera; and Xerox regained market share with its re-manufacturing program.

- External cost which also affects customer goodwill is diminished. Companies today have to take a "cradle-to-grave" approach of their products. Their actions today may haunt them in later years. In addition to punitive damages, top management may still be held personally liable. It is important to be socially responsible and adopt the correct course of action now, rather than *fight fires* later. Furthermore, these costs could be further diminished by working well with stakeholder teams. This would allow several options or alternatives to be considered in the product design and development stages, reduce cost, and quicken introduction into the market. Designing for the environment becomes a competitive weapon the firm can enjoy.

- The strategic importance of EMS decisions is very high. When poor environmental decisions are made, they are irreversible and the consequences are very high. The Union Carbide plant explosion in Bhopal, India in 1984 is still a glimpse reminder of the problems of poor environmental judgment. Apart from the so many deaths, irreversible blindness, long-lasting pollution, the ensuing public condemnation and labeling of the company has for long, dented its image and made it very difficult for the firm to operate. Poor environmental decisions lead to high production costs, poor public perception, poor quality, and would ultimately starve the firm of needed cash to conduct research and development or finance new projects.

## Environmental Action Box

### Hybrid Cars

The history of hybrid vehicles dates back to 1665. Flemish astronomer and Jesuit priest Ferdinand Verbiest developed plans for a miniature four-wheel unmanned steam "car" for Chinese Emperor

Khang Hsi. Since then, there have been several efforts to develop efficient hybrid cars. The most successful commercial effort to this effect could be traced to 1992 when Toyota Motor Corporation outlined in its document titled "Earth Charter," plans to develop and market vehicles that will yield the lowest emission possible. By 1997, Toyota began marketing Pirus in Japan and sold nearly 18,000 cars in its first year. By 1997-1999, other big auto manufacturers including Honda, GM and Ford introduced all-electric cars and those cars were sold mostly in California where environmental protection laws are stricter due to its poor quality air. Honda released its two-door Insight in 1999 and it is considered the first hybrid car to be offered to mass market in the United States. Insight was very successful winning several awards and was rated by EPA to receive 61 miles per gallon (mpg) in city driving and 70 mpg on highway. Toyota followed suit by releasing Toyota Pirus in the year 2000. Unlike the 2-door Insight, Pirus was the first 4-door sedan sold to mass market in the United States. By 2002, Honda introduced Honda Civic Hybrid and in 2004, Toyota Pirus II won Car of the Year Awards from Motor Trend Magazine and North American Auto Show. The demand for Pirus has skyrocketed in the US with Toyota's production of this vehicle rising from 36,000 to 47,000 in the US market. There was also a six month wait to purchase the vehicle in 2004. These hybrid cars are gradually making it into the mainstream market and other car makers have joined Honda and Toyota. The US auto manufacturer Ford introduced Escape Hybrid in September 2004 as both the first American hybrid and the first SUV hybrid.

The use of hybrid cars has enormous potential to minimize the burden on the environment. We shall summarize some of the implications on the environment that could be minimized through this effort. There has been a drastic increase in the demand for motor vehicles since the 1970s. It is estimated that there are over 700 million vehicles worldwide and if the present trend continues, this number could exceed 1 billion by 2025. There are more cars than adults in the US alone, vehicles are driven more than two trillion miles annually.

Motor vehicles contribute to all kinds of environmental and health hazards. While they are necessity for transportation purposes, however, the demand and use of vehicles need to be tamed to improve environmental quality. There are alternatives that could be pursued to limit the environmental impacts of vehicles.

*Global Warming* – About 5.5 million tons of carbon is released annually by the burning of gas, coal, and oil. The consequence is that heat is trapped in the atmosphere and causes the warming of the planet, thus producing greenhouse effect. Burning of gasoline contributes to greenhouse effect. When a gallon of gasoline weighing about 6 pounds is burned, it combines with oxygen to produce about 20 pounds of carbon dioxide ($CO_2$). With the increasing demand for motor vehicles, it is estimated that over 300 metric tons of carbon are produced by cars and light trucks every year in the US. The carbon dioxide emission from transportation constitutes more than a third of all other sources of emission. It is therefore prudent that efforts be spent on developing fuel efficient vehicles that would burn less fossil fuel or use alternative technologies to limit environmental burdens. The focus on hybrid vehicles could help achieve reduction in the volume of carbon that is generated.

The ensuing temperature fluctuations caused by Global warming could affect wildlife survival and the entire ecosystem.

In addition, carbon dioxide is the main greenhouse gas and minimizing its production could help to build a cleaner environment. The burning of fossil fuels by motor vehicles lead to the generation of several gases and pollutants which are environmentally unfriendly and unsafe to human health. We shall adapt and present some of the statistical information from *hybridcars.com* to show the significant gain of hybrid cars over conventional gas-powered vehicles. This information is presented in Table 1 below[7]:

---

[7]   The comparisons in Table 1 are based on 14,000 miles per year/EPA ratings.

| Table 1: Comparison of Conventional and Hybrid Cars | | | |
|---|---|---|---|
| **Gas/Pollutant** | **Car Type** | **Emission** | **Ecological/Health Risk** |
| **Carbon dioxide** | 2004 Toyota Camry 3.0L, 6 Automatic | 2004 Toyota Pirus 1.5L, 4CVT | Global warming; Severe disruption of global weather patterns. |
| | 11,000 pounds of carbon dioxide per year | 4,800 pounds of carbon dioxide per year | |
| **Carbon monoxide** | 2004 Cadillac SRX SUV 3.6L, 6 Automatic Bin 5 | 330 pounds of carbon monoxide per year | Poisonous gas that attacks the central nervous system |
| | 2005 Ford Escape Hybrid 2.3L, 4 CVT Bin 4 | 230 pounds of carbon monoxide per year | |
| **Nitrogen oxides** | 2004 Volkswagen Jetta 1.9L, 4 Automatic Bin 10 | 49 pounds of nitrogen oxide per year | Global warming, formation of ground-level ozone, acid rain, and smog; Respiratory problems |
| | 2004 Honda Civic Hybrid 1.3 L, 4 CVT Bin 9 | 17 pounds of nitrogen oxide per year | |
| **Particulate matter** | 2004 Range Rover 4.4L, 8 Automatic Tier 1 | 670 grams of particulate matter per year | Consists of particles of smoke, soot and dust. Health hazard especially lungs and bloodstream. |
| | 2005 Honda Accord Hybrid 3.0L, Automatic ULEV II | 240 grams of particulate matter per year | |
| **Hydrocarbons** | 2004 Hummer H2 6.0L, 8 Automatic HDT-Bin 11 | 29 pounds of hydrocarbons per year | Air toxicity, smog, lung and tissue diseases, birth defects |
| | 2004 Honda Insight 1.0L, 3 Manual Bin 9 | 8 pounds of hydrocarbons per year | |

In addition to gases and pollutants associated with motor vehicles, there are also the production of sulphur oxides and lead. Sulphur oxides contribute in the formation of acid rain and are potentially dangerous to children and elderly people. The production of lead is also of concern although some countries have banned the use of leaded gasoline. Lead poisoning is known to damage organs and other tissues in the body.

This table did show that a shift to hybrid vehicles could help to minimize environmental burden by improving fuel efficiency and thereby reducing the amount of pollutants that are produced. As we mentioned earlier in this chapter, some countries now give tax incentives to encourage the shift to hybrid vehicles. Even with this shift, motor vehicles in general pose major environmental hazard. There are still the issues of water pollution from runoff oil or fluids or chemicals that manage to sip into the waterways; with the increasing number of vehicles on the road, noise pollution is still a major concern; and there is concern about solid waste disposal. All these problems continue to affect both ecological and health risks. Motor vehicles need to be designed more efficiently, made lighter, and resort to alternative energy sources to reduce environmental burden. Hybrid vehicles are contributing in improving environmental quality but human conscious effort to use mass transit systems and change driving habits could even help further. Countries like the United States also offer tax incentives to commuters who use mass transit systems to commute to work.

# Chapter 2

# Sustainable Manufacturing

Earth's resources are limited. With the explosion in world population and the increasing rate of consumption, it will be increasingly difficult to sustain the quality of life on earth if serious efforts are not made now to conserve and effectively use the earth's limited resources. It is projected that the current world population of 5.6 billion people would rise to 8.3 billion people by the year 2025 [Furukawa 1996]. This is an increase of 48.21% from the current level. Yet, earth's resources such as fossil fuels, landfills, quality air and water are increasingly being depleted or polluted. So, while there is a population growth, there is a decline in the necessary resources to sustain the increasing population. Since the mid-1980s, we have witnessed a rapid proliferation of new products with shorter life cycles. This has created tremendous wastes that have become problematic as more and more of the landfills are usurped. Increasingly, more and more environmental activist groups are forming and with consumer supports, are putting pressures on corporations to improve their environmental performance. These efforts are also being supported by the increase in the number of new legislatures to protect the natural environment. Thus, responsible manufacturing is needed to achieve sustainable economic development. Strikingly, studies have linked economic growth to environmental pollution [Madu 1999]. Thus, there is a vicious cycle between improved economic development and environmental pollution. This traditional belief in a link between environment pollution and economic growth often is a hindrance to efforts to achieve sustainable development. Sustainable manufacturing is therefore, a responsible manufacturing strategy that is cognizant of the need to protect the environment from environmental pollution and degradation by conserving the earth's limited resources and effectively planning for the optimal use of resources and safe disposal of wastes. In the past, manufacturers have been lukewarm about any strategy to develop sustainable manufacturing. They viewed such strategies as expensive and not economically viable. However, this mood is gradually changing as more and more big companies are developing environmentally conscious manufacturing strategies through their entire supply chain. Many have also seen that environmentally conscious manufacturing can become an effective competitive strategy. Thus, sustainable manufacturing makes wise business sense and can lead to improved bottom-line. We shall in this chapter, trace the origins of sustainable development, which gave rise to sustainable manufacturing. Further, we shall identify different strategies to sustainable manufacturing and then present cases of successful implementation of sustainable manufacturing by multinational corporations such as Kodak and Xerox.

## Sustainable Development

The origins of sustainable development can be traced to the United Nations publication in 1987 titled the Brundtland Report. This report is named after Mrs. Brundtland, Prime Minister of Norway who chaired the UN World Commission on Environment and Development. The report focused on the problems of environmental degradation and states that "the challenge faced by all is to achieve sustainable world economy where the needs of the entire world's people are met without compromising the ability of future generations to meet their needs." This report received an international acclaim as more and more people are concerned with the theme of the report on environmental degradation. Since its publication, the world community has convened several conferences on how to achieve sustainable development. In 1992, the UN organized the Earth Summit in Rio de Janeiro, Brazil with a focus on how to get the world community to cut down on the use of nonrenewable resources in other to achieve sustainable development. This conference highlighted the disparate views between the industrialized and the developing countries on how sustainable development could be achieved with those from the Southern Hemisphere seeing dependence on the use of natural resources as a prerequisite to their economic growth. Several publications have emerged on sustainable development since the conference.

Duncan [1992] defined sustainable development as an "economic policy which teaches that society can make the appropriate allocation of resources between environmental maintenance, consumption, and investment." However, such balance is difficult to achieve when a nation becomes completely dependent on the exploitation of natural resources to satisfy its social and economic needs [Madu 1996]. Furthermore, with the absence of a developed private sector, countries faced with harsh economic realities such as poverty and over population, are more likely to focus on exploitation of natural resources and deployment of inappropriate technologies for manufacturing. Such attempts may hinder the global efforts to achieve sustainable development. Following this debate, Singer [1992] argues that sustainable development is akin to a "New Economic Order" that may not encourage reasonable and realistic development from the Southern Hemisphere. Rather, it could be seen as an attempt to make the South financially dependent on the North. This he refers to as a Robin Hood effect which may result in the transfer of funds from the poor in the rich countries to the rich in the poor countries. Clearly, achieving sustainable development is a goal for the entire world otherwise; marginal efforts by each country will be ineffective. Fukukawa [1992] pointed out that "current global environmental problems may bring about a crisis that could never have been anticipated by our predecessors. Since the very inception of history, humankind has been pursuing technological development to protect itself from the threats and constraints of nature. However, economic activities triggered by these technological developments have grown large enough to destroy our vital ecosystem." This view is shared by many around the world and has been a motivating force in seeking for responsible manufacturing through sustainable manufacturing. While many companies in the industrialized countries have embarked on the road to sustainable development, it is important to achieve environmental conformance throughout the world. After all, noncompliance may affect the supply chain especially since some of the raw materials may be generated from the poorer nations. Getting these countries to participate in sustainable development will requires understanding their perspectives on economic and social development and how they could be assisted by the more affluent nations. The problems in developing countries are better explained by Kamal Nath, India's then Minister of the Environment, when he noted in the Rio de Janeiro conference that "Developed countries are mainly responsible for global environmental degradation and they must take the necessary corrective steps by modifying consumption patterns and lifestyles; developing countries can participate in global action, but not at the cost of their development efforts ... On climate change and greenhouse gases, India's stand is that global warming is not caused by emissions of the gases per se but by excessive emissions. The responsibility for cutting back on emissions rests on countries whose per capita consumption is high. India's stand is that emission in developed countries be reduced to tally with the per capita emission levels of developing countries." This view obviously, is controversial in industrialized countries. However, what it points out is the link or the perception of a link between environmental pollution and economic growth. In fact, as Figure 2.1 shows, the emission levels of carbon tend to support such a link. This figure, suggests a direct relationship between carbon emission and economic growth when the cases for OECD (Organization of Economic Cooperation and Development) are compared to the cases for non-OECD nations.

In 1997, the UN conference on Climate Change was held in Kyoto, Japan. This conference further raised some doubts and disagreements

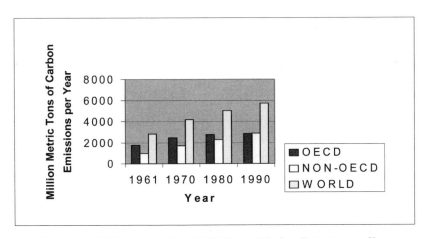

*Figure 2.1: Million Metric Tons (2.204 lbs.) of Carbon Emissions per Year*

between member nations, non-governmental organizations, labor unions, and environmental activists. A preamble placed on the Internet states as follows, "The threat of global warming has brought more than 140 governments together in intensive negotiations to try to limit the emission of carbon dioxide and other greenhouse gases that trap heat in the atmosphere. But history, geography, economics and politics are driving them apart. Island states fear the rising oceans that warming may cause. Oil producers fear what lessening the world's dependence on fossil fuels would mean to them. Big industrial nations worry that emission limits might slow their economies. Poor nations say they should not have to bear the same burden as the rich." Obviously, sustainable development is intertwined with politics and economics and these may impede the effort to achieve sustainable development. Strategies to achieve this goal must therefore, take into consideration these concerns. Clearly, sustainable development cannot be achieved without sustainable manufacturing. Sustainable manufacturing is one of the processes or strategies to achieve the goal of sustainable development.

Sustainable manufacturing as a strategy will require the re-engineering of the organization to change design, process, work attitudes and perceptions. It requires the entire organization to be environmentally conscious and will require the support and participation of top management. More importantly, it will require investment in the future and retraining of the work force. Sustainable manufacturing is a capital venture that a company must undertake and this is a risk that some may not yet be ready for especially from the developing economies. Yet, from all indications, those corporations that have embarked on this bold step are reporting dramatic successes as we shall outline later.

The Brundtland Report was instrumental in getting the world focus on sustainable development. However, the formation of the Business Charter for Sustainable Development (BCSD) by a group of 50 business executives provided the momentum for much of businesses involvement in sustainable manufacturing. BCSD was formed in 1990 in preparation of Business activities at UNCED. This group was headed by Stefan Schmidheiny and published a book titled "Changing Course." This book detailed with case studies, challenges facing business in a sustainable environment. In January 1995, BCSD merged with another influential group with strong business ties known as the World Industry Council for the Environment (WICE). WICE is an initiative of the International Chamber of Commerce (ICC) based in Paris while BCSD was based in Geneva. These two groups shared common goals and attracted executives from similar organizations although BCSD was an executive-based group. The result of this merge is the World Business Council for Sustainable Development (WBCSD). WBCSD is presently, a coalition of 125 international companies that share a commitment to environmental protection and to the principles of economic growth through sustainable development. Its membership is drawn from 30 countries and more than 20 major industrial sectors. The aims of WBCSD as listed in its web page are stated below as follows (http://www.wbcsd.ch/whatis.htm):

- Business leadership—To be the leading business advocate on issues connected with the environment and sustainable development;
- Policy development—To participate in policy development in order to create a framework that allows business to contribute effectively to sustainable development;
- Best practice—To demonstrate progress in environmental and resource management in business and to share leading-edge practices among our members;
- Global outreach—To contribute through our global network to a sustainable future for developing nations and nations in transition.

The participation and support of many executives and major industrial sectors in sustainable development issues gave the momentum to corporate focus on sustainable manufacturing or environmentally conscious manufacturing.

Schmidheiny in his 1992 article discusses the term ecoefficiency. He defines it as "companies which add the most value with the least use of resources and the least pollution." This definition clearly linked industrial production to achieving sustainable development and shows that ecoefficiency or sustainable development can be achieved only when limited natural resources are optimized and environmental waste and pollution are minimized. Thus, corporate responsibility for sustainable development is obvious and corporations and their executives by participating in WBCSD are leading the way to achieving sustainable manufacturing. Sustainable manufacturing is therefore, synonymous to ecoefficiency. We shall therefore, define sustainable manufacturing as a means for manufacturers to add the most value to their products and services by making the most efficient use of earth's limited resources, generating the least pollution to the environment, and targeting for environmental clean production systems. Although we emphasize sustainable manufacturing, it should be apparent that the goal of environmental clean production cannot be achieved if the service component of the manufacturing system is not environmentally conscious. The service sector must contribute by ensuring that its services are environmentally efficient. For example, can the purchasing and receiving department conserve its use of paper for placing orders? Obviously, such a simple case can be achieved by using recycled papers and packaging, and by placing most orders through interconnected computers or the Internet in a paper-less environment. Thus, our focus is on both the manufacturing and service sector working in harmony to achieve the goal of environmentally conscious manufacturing.

## Strategies for Sustainable Manufacturing

Several strategies have been developed to achieve sustainable manufacturing. We shall briefly discuss the different strategies. The aim of each of these strategies is to find a better way to make more efficient use of the earth's limited resources, minimize pollution and waste. Some of these strategies may appear in more details in subsequent chapters.

- Inverse Manufacturing—This strategy is based on prolonging the life of a product and its constituent components. Umeda [1995] refers to this as a closed-loop product life cycle. Simply stated, the life of any product can be extended by disassembling the original product at the end of its original life into components that could be reused, recycled, maintained or up-graded. Focus is on limiting the amount of components that are disposed or discarded as wastes. When this is done, environmental costs are minimized [Yoshikama 1996]. Inverse manufacturing gets its name from the reverse approach to recovery of the components that make up a product. Due attention is given at the conception of the product to the ease of disassembly. This will make it possible to reclaim component parts for future use thereby prolonging the life of the product. There are many examples of inverse manufacturing. For example, older computers are frequently upgraded to give them more capabilities. By retaining much of the computer unit and adding only the needed features, its life is further extended. Also, important precious metals present in some older computer units such as silver, platinum and gold can be extracted and reused in building newer models when it is no longer economical or feasible to upgrade the unit. These activities reduce waste through recovery, recycling and reuse of materials. In the paper industry also, the use of recycled paper rather than virgin pulp in new paper production prolongs the life of the original virgin pulp. Inverse manufacturing has obvious advantages in extending the life of the product, minimizing waste of materials and conserving the landfills. The goal however, should be to keep waste to a bare minimum.

- Recycling—Recycling is one of the better-known strategies for sustainable manufacturing. In most communities, it is mandatory to participate in recycling programs. Many people identify with recycling of newspapers, packages, soda cans, bottles, and in fact, are required to separate them from other garbage for recycling purposes. Although there are arguments about the weaknesses of the current day recycling policies, however, the aim of recycling is to focus our attention on the finite resources available to mankind. The earth is composed of about 30% land and the rest is water. Our landfills are gradually filling up. If we continue to discard and dump wastes, the landfills will be filled up. We depend on the limited earth's resources for economic growth and if we are not able to thoroughly recycle and extend the lives of these resources, the future will be blink. Thus, a recycling policy that is efficient is needed. Such policy should be efficient and encourage more people and industries to participate in the program.

- Re-manufacturing—is the process of rebuilding a unit or machinery to restore its condition to "as good as new." This may involve reuse of existing components after overhaul, replacement of some component parts, and quality control to ensure that the remanufactured product will meet new product's tolerances and capabilities. The remanufactured product will normally come with a new product warranty. To make remanufacturing effective, the following steps are normally taken:

  1. Collection of used items—This could be achieved through a recycling program where used or expired original products are collected from the customer and reshipped to the manufacturer. Some examples of these are drum and toner cartridges for computer printers and photocopying machines, auto parts, etc.
  2. These items on receipt are inspected based on their material condition and a determination can be made on the economic feasibility of remanufacturing them.
  3. Subsequently, the items are disassembled. If the full unit cannot be remanufactured, some components may be recovered for use in other components. Otherwise, the original item can be restored to a condition as good as new through repair and servicing. The recovery process must be efficient and focus on strategies that are conducive to the environment.

It is important that new products are designed for ease of disassemble and recovery of parts. This will make it more economical to conduct remanufacturing activities since it will be easier to determine which parts need repair or replacement. This will also help in effective planning of the master production scheduling by minimizing the production planning time and parts inventory levels.

- Reverse logistics—requires that manufacturers take a "cradle-to-grave" approach of their products. This management of a product through its life cycle does not end with the transfer of ownership to the consumer and the expiration of warranty. Rather, the manufacturer is forever, responsible for the product. This is often referred to as "product stewardship." [Dillion and Baram 1991]. Roy and Whelan [1992] noted that this is a "systematic company efforts to reduce risks to health and the environment over all the significant segments of a product life cycle." Product stewardship is driven by public outcry about the degradation of the environment. This has led to new legislatures making manufacturers responsible for the residual effects of their product on the environment with no time limit. As a result, more and more companies are responding by developing environmentally responsible strategies. Some are also seeing that such strategies are good for business and may lead to competitive advantages. The concept of product stewardship as outlined by Roy and Whelan [1992] requires a focus on the following:

1. Recycling
2. Evaluation of equipment design and material selection
3. Environmental impact assessment of all manufacturing processes
4. Logistics analysis for the collection of products at the end of their lives
5. Safe disposal of hazardous wastes and unusable components
6. Communication with external organizations—consumer groups legislature, and the industry at large.

This focus is embodied in the reverse logistic strategy. It is a new way for manufacturers to view their products and develop a business model that could enable them to profit from developing a product stewardship approach. Obviously, by using remanufacturing strategy, the manufacturer can save significantly from the cost of labor and materials. Giuntini [1997]. Note that about 10 to 15% of the gross domestic product could be affected, by adopting reverse logistic as a business strategy. Furthermore, about 50 to 70 percent of the original value of an impaired material can be recovered from customers. In addition, the cost of sales (direct labor, direct material, and overhead) which currently, averages 65 percent to 75 percent of the total cost structure of a manufacturer can be reduced by as much as 30 percent to 50 percent through reverse logistics. He identified the by-products of reverse logistics as follows:

- Industrial waste throughout the manufacturing supply chain, would be reduced by as much as 30 percent
- Industrial energy consumption would be noticeably reduced
- Traditionally under-funded environmental and product liability costs would be better controlled and understood.

He suggested the following 10 steps for a manufacturer to implement a reverse logistics business strategy:

- Products must be designed for ease of renewal, high reliability, and high residual value.
- Financial functions must be restructured to cope with different cash-flow requirements and significant changes in managerial accounting cues.
- Marketing must reconfigure its pricing and distribution channels.
- Product support services and physical asset condition monitoring management systems must be implemented to manage manufacturer-owned products at customer sites.
- Customer order management systems must be implemented to recognize the need for the return of an impaired asset from a customer site.
- Physical recovery management systems must be implemented to manage the return of impaired physical assets.
- Material requirements planning management systems must be implemented to optimize the steps required to be taken upon the receipt of recovered impaired assets.

- Renewal operational processes must be established to add value to impaired assets.
- Re-entry operational processes must be established to utilize renewed assets.
- Removal processes must be established to manage nonrenewed assets.

- Eco-labeling—The aim of eco-labeling is to make consumers aware of the health and environmental impacts of products they use. It is expected that consumers will make the right decision and choose products that will have less environmental and health risks. By appropriately labeling the product and providing adequate product information for consumers to make the choice between alternative products, it is hoped that manufacturers will move towards developing environmentally conscious production systems. Eco-labeling as a strategy is therefore, intended to identify the green products in each product category. It could be perceived as a marketing strategy that is partly driven by legislatures and partly driven by consumers concern for the degradation of the environment. Many of the eco-labeling schemes are based on the life cycle assessment (LCA) of a product and take the "cradle-to-grave" approach by evaluating the environmental impacts of the product from the extraction of the raw material to the end of the product's useful life. However, some of the popular eco-labeling schemes do not take this approach. The German "Blue Angel" mark which is one of the best known eco-labeling schemes focuses on the environmental impacts of the product at disposal and the Japanese EcoMark focus on the contributions of the product to recycling Using Eco-labeling, (http://www.uia.org/uiademo/str/v0923.htm).

Eco-labeling is increasingly being used in many industries and consumers are paying attention as opinion polls tend to suggest [Using eco labeling, 1999]. However, for eco-labeling to be effective, the public needs to be well informed and the labeling scheme must be credible. As has been suggested, it is important that all the major stakeholders (i.e., consumers, environmental interest groups, and producers) participate in developing the eco-labeling schemes. Also, information presented on the content of the product has to be valuable and understandable to consumers. There is a need for a standardized scheme in each product category to make it easier for comparative judgments. One of the major problems facing eco-labeling schemes is that it is voluntary and often, administered by third parties. Bach [1998] argued that mandatory eco-labeling schemes would be illegal within the context of the World Trade Organization and act as a barrier against international trade. He is of the opinion that regulatory measures will not reduce environmental degradation and further note that different countries have different environmental policies and standards as well as different economic policies and standards.

However, market forces and not government laws and legislatures drive eco-labeling. We operate in a global environment and without a standardized eco-labeling scheme; the entire supply chain will be affected. It is clear that many producers in industrialized countries source their raw materials and parts from different countries. If a standardized eco-labeling scheme is not developed, the entire supply chain will be affected and it will be difficult to implement an eco-labeling scheme that is based on a cradle-to-grave approach. Furthermore, the changes we have observed in the market economy since the 1980s as a result of the total quality movement (TQM) and the subsequent development of the ISO 9000 series of product standards suggest that international standards on eco-labeling are not far from implementation. In fact, with the success of ISO 9000, the International Organization for Standards (ISO) has developed the ISO 14000 series of standards with a focus on guidelines and principles of environmental management systems. The technical committee (TC 207) charged with developing standards for global environmental management systems and tools, has environmental labeling as part of its focus. ISO 14020 deals with the general principles for all environmental labels and declarations [Madu 1998]. As expected, these standards will be widely adopted and when that happens, businesses will be expected to follow accordingly in order to compete in global markets. ISO already has classifications for eco-labeling schemes and the Type I eco-labels have the greatest impact on international trades. A third party to products that meet specified eco-labeling criteria grants certification. The issue is not to have each country develop its own plan for eco-labeling but, for world bodies such as ISO to institute a standardized scheme that will be cognizant of the limitations poorer nations may face. Indeed, ISO has four standards dealing with eco-labeling. These are ISO 14020, ISO 14021, and ISO 14024 and ISO 14025. Although ISO standards are voluntary, with the worldwide acceptance of ISO, it is expected that many companies and countries will work within the guidelines of these standards. Environmental protection should be a worldwide effort and without such an effort, the whole idea will be marginalized. Finally, some have argued that eco-labels do not boost sales [Christensen 1998] but it is too early to verify this claim since the public has to be sufficiently aware. Also, sales should not be the single criterion for environmental protection. Due concern should be given to the consumer's need to be aware of the content of the product and having the ability to make a purchasing decision based on that information.

- ISO 14000—is a series of international standards on environmental management. These standards are being put up by the International Organization for Standards (ISO) with the objective to meet the needs of business, industry, governments, non-governmental organizations and consumers in the field of the environment. These standards are voluntary; however, they continue to receive the great support of ISO member countries and corporations that do business in those countries. We shall not go into the details of these standards since ISO 14000 is a chapter in this book. We shall however, present a table that lists the ISO 14000 standards and other working documents at the time of writing. This is to help draw your attention to the work done by ISO on environmental management. However, the work of the ISO technical committee working on ISO 14000 family of standards is to address the following areas:

Environmental management systems.

Environmental auditing and other related environmental
investigations

Environmental performance evaluations.

Environmental labeling.

Life cycle assessment.

Environmental aspects in product standards.

Terms and definitions.

Table 2.1 shows the listing of approved standards and drafts at their different stages of development as of 1999. Later on in Chapter 4, we shall present an updated version of these standards which was last revised on November, 2004.

*Table 2.1: ISO 14000 Family of Standards and Ongoing Work*

| Designation | Publication | Title |
|---|---|---|
| ISO 14001 | 1996 | Environmental management system—Specification with guidance for use |
| ISO 14004 | 1996 | Environmental management system—General guidelines on principles, systems and supporting techniques |
| ISO 14010 | 1996 | Guidelines for environmental auditing—General principles |
| ISO 14011 | 1996 | Guidelines for environmental auditing—Audit procedures—Auditing of environmental management systems |
| ISO 14012 | 1996 | Guidelines for environmental auditing—Qualification criteria for environmental auditors |
| ISO/WD 14015 | To be determined | Environmental assessment of sites and entities |
| ISO 14020 | 1998 | Environmental labels and declarations—General principles |
| ISO/DIS 14021 | 1999 | Environmental labels and declarations—Self declared environmental claims |
| ISO/FDIS 14024 | 1998 | Environmental labels and declarations—Type I environmental labeling—Principles and procedures |

*Table 2.1: (Continued)*

| Designation | Publication | Title |
|---|---|---|
| ISO/WD/TR 14025 | To be determined | Environmental labels and declarations—Type III environmental declarations—Guiding principles and procedures |
| ISO/DIS 14031 | 1999 | Environmental management—Environmental performance evaluation—Guidelines |
| ISO/TR 14032 | 1999 | Environmental management—Environmental performance evaluation—Case studies illustrating the use of ISO 14031 |
| ISO 14040 | 1997 | Environmental management—Life cycle assessment—Principles and framework |
| ISO 14041 | 1998 | Environmental management—Life cycle assessment—Goal and scope definition and inventory analysis |
| ISO/CD 14042 | 1999 | Environmental management—Life cycle assessment—Life cycle impact assessment |
| ISO/DIS 14043 | 1999 | Environmental management—Life cycle assessment—Life cycle interpretation |
| ISO/TR 14048 | 1999 | Environmental management—Life cycle assessment—Life cycle assessment data documentation format |
| ISO/TR 14049 | 1999 | Environmental management—Life cycle assessment—Examples for the application of ISO 14041 |
| ISO 14050 | 1998 | Environmental management—Vocabulary |
| ISO/TR 14061 | 1998 | Information to assist forestry organizations in the use of the Environmental Management Systems standards ISO 14001 and ISO 14004 |
| ISO Guide 64 | 1997 | Guide for the inclusion of environmental aspects in product standards |

**Notes:**

CD = Committee Draft;
TR = Technical Report;
DIS = Draft International Standard;
FDIS = Final Draft International Standard;

*Source: Adopted from "ISO 14000—Meet the whole family!" retrieved 3/11/1999 from http://www.tc207.org/home/index.html.*

- Life cycle assessment—we shall adopt the definition provided by ISO for life cycle assessment (LCA). It is defined as "a technique for assessing the environmental aspects and potential impacts associated (with products and services)... LCA can assist in identifying opportunities to improve the environmental aspects of (products and services) at various points in their life cycles." This concept is often referred to as the "cradle-to-grave" approach. It requires that emphasis be placed on the environmental impacts of production or service activities from the product conception stage (i.e., raw material generation) to the end of the product's life (i.e., recovery, retirement or disposal of the product). Thus, the manufacturer is responsible for the environmental impacts of the product through different stages in its life cycle. Life cycle assessment often involve three major activities [Affisco 1998]:

1. Inventory analysis—this deals with the identification and quantification of energy and resource use as well as environmental discharges to air, water and land.
2. Impact analysis—is a technical assessment of environmental risks and degradation.
3. Improvement analysis—identifies opportunities for environmental performance improvement.

Notice also that several of the ISO standards listed in Table 1 deal with Life Cycle Assessment. Already, ISO 14040 on principles and framework and ISO 14041 on goal and scope definition and inventory analysis have been adopted as standards.

- Design for the environment—consequent to the growing demand for improvement in environmental performance is the growing need to change the traditional approach to designing. This strategy calls for an efficient designing of products for environmental management. Products are to be designed with ease of disassembly and recovery of valuable parts. Such design strategies will conserve energy and resources while minimizing waste. In designing for the environment, tradeoffs are made between the different environmental improvements over the product life cycle. Three main design strategies are design for recyclability; design for remanufacture; and design for disposal.

1. Design for recyclability—This involves the ease with which a product can be disassembled and component parts recovered for future use. For example, with computer units, precious metals can be easily recovered for use in new computers. For chemical compounds, the focus is on separability of materials to avoid contamination and waste of energy in recovering these materials.

2. Design for Remanufacture—This recognizes the different stages of equipment or product wears. For example, certain parts of machinery (i.e., auto parts) could be recovered, remanufactured and restored to a state as good as new. Reusing them in newer products could further extend the lives of such parts. The challenge is how to design the original product for ease of recovery of those parts. We notice for example that newer computer systems are designed with the ease of upgrading them. Thus, new capabilities could be added to the system without having to dispose of the old unit.

3. Design for disposability—This recognizes the fact that many of the earth's landfills are filling up at an alarming rate. Further, many of the deposits are hazardous and unsafe. It is important to design the product with the ease of recycling and disposal. The final waste generated from the product should also be disposed safely.

## Environmental Action Box

The case studies presented here are some of the popular success stories from leading manufacturers to show that responsible design; production and packaging that are environmentally sensitive are profitable. Many of these companies have witnessed growth in sales and revenue and attribute these successes to their environmental management programs.

### *Kodak Single-Use Camera:*

The Kodak single-use camera (SUC) is perhaps, one of the most remarkable successes stories. Kodak first introduced this product in the U.S. in 1987 and it is now, the company's fastest growing product category. This product is now the company's centerpiece in its efforts in recycling, re-use, and product stewardship. Interestingly, the single-use camera was introduced as an inexpensive camera and not as an environmental product. It became known widely as a disposable camera and was even dubbed an environmental "ugly duckling." However, through innovation, commitment, and hard work, Kodak has transformed this product into an environmental success story. How did Kodak achieve this feat?

The product was designed for the environment and it is dubbed by some as the best example of closed-loop recycling. The recycling of Kodak single-use camera is a three-prong process that involves the active participation of photofinishers and a strategic partnership with other SUC manufacturers. In fact, Kodak credits photofinishers with most of the success achieved in recycling and reusing Kodak's SUCs. The new SUCs use 20 percent fewer parts from the design features to the actual film processing stage. Photofinishers return the camera after processing to Kodak and are reimbursed for each camera returned plus shipping cost. In the U.S., a 63 percent return rate has been achieved for recycling. This is equivalent to fifty million SUCs or enough SUCs to fill up 549 tractor-trailer loads.

The three-prong recycling process as detailed by Kodak in its website

http://www.kodak.com/US/en/corp/environment/performance/recycling
/suc.shtml, retrieved 3/2/99) is as follows:

1. Photofinishers ship the SUCs to three collection facilities around
   the world and Kodak maintains a recycling program in more than
   20 countries. Through the strategic partnership with other SUC
   manufacturers such as Fuji, Konica, and others, they jointly
   accept each other's products even though Kodak cameras are in
   the majority. The products are sorted according to manufacturer
   and camera model. These cameras arrive at these facilities in
   recyclable cardboard.
2. Kodak cameras are shipped to a subcontractor facility for
   processing. The packaging is removed and any batteries in the
   camera are recovered. The camera is cleaned up and undergoes
   visual inspection. The process of remanufacturing the SUC has
   begun and the old viewfinders and lenses are replaced. New
   batteries are also inserted. Those parts that could be reused are
   retained after rigorous quality control checks.
3. The SUCs are now shipped to one of Kodak's three SUC
   manufacturing plants. This is the final assembly where new
   packaging made from recycled materials (with 35% post-
   consumer content) is added. The camera is now ready for use.

There are some lessons that could be learned from the Kodak
experience:

1. Kodak notes that by weight, 77 to 86 percent of Kodak SUCs can
   be recycled or re-used. Yet, these products maintain high quality,
   attract huge demand and are profitable. This suggests that
   responsible environmental programs can be competitive and help
   the firm to achieve its profit motives. Recycling programs such as
   this can help the manufacturer to save significantly by cutting
   down the cost of material, labor, and to achieve faster response to
   the market. It is estimated that it takes about 30 days from the
   time of collection of an SUC to reclamation and re-introduction to
   the market.

2.  Strategic partnership and working with vendors may be instrumental in effective recycling programs. Photofinishers have an incentive to participate and the cost of recycling can be shared through industry partnership as demonstrated in the case of Kodak's SUC.

3.  Design for the environment is essential. Products for recycling, reuse, and remanufacturing must be designed for ease of disassemble and recycle. For example, with the SUC, it is easy to reclaim the packaging and recover component parts. Such design cuts down on cost and therefore, makes designing for the environment attractive.

4.  Conservation of resources is achieved through effective recycling programs. For example, the equivalent of 549 tractor-trailer loads of SUCs has been recycled. This is said to be equivalent to 3,333 miles of cameras laid end-to-end. Imagine the enormous pollution this will create if these cameras are "disposables." How much landfill space will be needed to contain them? Since the recycling and reuse program began in 1990, more than 200 million cameras have been recycled. Since 1990, there has been an exponential growth in the number of cameras recycled with the number increasing from 42.1 million in 1996 to 51.9 million in 1997. This is shown in the figure below from the data presented by Kodak.

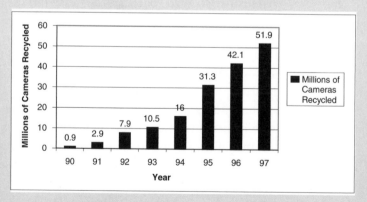

*Figure 2.2: Millions of Kodak Single Use Cameras Recycled. Adopted from, http://www.kodak.com/US/en/corp/environment/performance/recycling/suc.shtml*

5. The Kodak SUC is an example of a cradle-to-grave approach of a product. The manufacturer designs the product so that it has control over the life of the product. This is achieved by ensuring that the consumer ultimately, will return the camera to a photofinisher for processing and the photofinisher is given an incentive to participate in the environmental management program. The manufacturer takes the responsibility of disassemble and reuse and disposal of the product. This process ensures an effective environmental program that makes the manufacturer responsible for the life of the product.

The Kodak SUC is a success story that has helped reduce environmental degradation and has achieved tremendous economic success. Next we look at the environmental program at Xerox.

### *Xerox:*

Xerox has a long history of developing sustainable products that dates back to 1967. Its strategy involves design for environment and life cycle product valuation. In 1967, the company embarked on a metal recovery program from photoreceptor drums and continues today to reclaim metals for reuse or remanufacturing purposes. Its design strategy today is known as "Waste-Free" design. How does this program work? Machines are recovered from customers through trade-ins and lease options. Many of the components of the xerographic machines that can still perform at their original specifications are recovered for reuse and remanufacturing. In 1997 alone, more than 30,000 tons of returned machines were used to remanufacture new equipment. Within the past five years, Xerox has more than doubled the number of machines it remanufactures. The remanufactured machines still meet Xerox's strict quality guidelines and are offered with the same Xerox Total Satisfaction Guarantee. These machines are designed for ease of disassemble, and Xerox takes the responsibility of the product's life cycle. As a result of the company's environmental efforts, natural resources are conserved and new machines are designed with fewer replacement parts.

Xerox works with its customers to carry out the recycling program. Customers of copy cartridges are provided with prepaid return labels that enable them to reuse the packaging from the new cartridge to ship the used cartridges to Xerox. The reused cartridges are then remanufactured. In 1997, Xerox achieved a return rate of 65% for print and copy cartridges. This is now the industry benchmark. Xerox also maintains a Waste Toner Return Program. This program allows customers to return waste toners for remanufacturing, reuse and recycle. This program is credited with the recovery of millions of pounds of toner which would have otherwise, been sent to landfill.

Xerox adopts a company-wide environmental program that tracks its product's life cycle and ensures environmental protection. Its recycling program works well because of the extensive network of people who participate in the delivery process to monitor the environmental and other potential impacts of the product on Xerox. A framework of its successful recycling program is shown below:

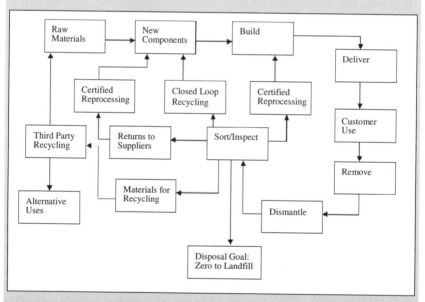

*Figure 2.3: Xerox Recycling Program. Adapted from "Sustainable product development—The First 30 years," http://www.xerox.com/ehs/1997/sustain.htm, page 8, retrieved 3/2/99.*

Notice that similar to the case of Kodak Single Use Camera, the Xerox recycling program adopts a closed-loop approach. This ensures that no waste is incurred as the material is continuously recycled, reused and remanufactured. In the end, the goal of achieving zero disposals to the landfill can be achieved. However, this can only be feasible if the product is designed for the environment. In other words, recycling will become a profitable and an economical alternative to waste.

The approaches that Kodak and Xerox are taking are innovative in that both manufacturers develop effective collection systems for their expired products. Many of the existing recycling programs depend on the garbage industry and municipalities rather than on vendors and suppliers who have stakes in the recycling process. The lack of a well-designed recycling program has created displeasure and dissatisfaction that often times denigrate the value of recycling programs.

Some of the achievements of the Xerox program are outlined below http://www.xerox.com/ehs/1997/iso.htm page 2:

1   There has been a dramatic reduction in the amount of hazardous waste generated and 1997 saw a reduction of about 20% from the 1996 level.
2   There is a reduction in the amount of solvents sent off-site for recycle by about 13% in 1997.
3   Non-hazardous solid wastes designated for the landfills dropped by 20% in 1997.
4   Recycling rate of non-hazardous waste increased to 85%.
5   There is an increase in monetary savings from waste-free initiatives.
6   SARA (Superfund Amendment and Reauthorization Acts) air emission levels remain unchanged.

To add to Xerox environmental program, many of its processes are ISO 14001 certified. Xerox noted its ISO 14001 results as

•   Lowered energy consumption.
•   Increased recycling of various materials, including metals, plastics and cardboard.

- Reduced waste to landfill.
- Improved procedures for managing hazardous waste.
- Development of an electronic document management application for customers implementing their own environmental management systems.

## Conclusion

In this chapter, we have traced the origins of sustainable development to the United Nations publication in 1987 titled the Brundtland Report. This report provided further impetus for the world community to focus on environmental protection. Since this report was presented, several conferences have been organized by several world agencies such as the United Nations to focus the world attention to the need of pollution prevention and resource conservation. However, it was the formation of the Business Charter for Sustainable Development (BCSD) by a group of 50 business executives that provided the momentum for much of businesses involvement in sustainable manufacturing. As we mentioned earlier, the publication of a book "Changing Course" by BCSD, outlined the challenges facing business in a sustainable environment. The merging of BCSD with the World Industry Council for the Environment (WICE) further expanded the industry interests in sustainable manufacturing. These two groups shared common goals and attracted several executives and industries from around the world. The new group now known as the World Business Council for Sustainable Development (WBCSD) is presently, a coalition of 125 international companies that share a commitment to environmental protection and to the principles of economic growth through sustainable development. Its membership is drawn from more than 30 countries and from more than 20 major industrial sectors.

The interest of businesses and the world community at large has spurned a lot of interests on environmental protection since the 1990s. Subsequently, several strategies have been developed to deal with environmental protection issues. We have outlined some of these strategies in this chapter. Environmental protection is increasingly seen as a competitive tool. Companies now pride themselves for their continuous efforts to protect the environment and provide consumers with environmentally clean products. We continue to see this effort expanded to include not only the content of the product but also its tertiary value such as packaging and preparation of the order. Also, the entire supply chain is now involved with environmental protection efforts. Sustainable manufacturing is in vogue now and every progressive company should have a strategy to achieve its environmental protection goals. It is important to know that with many nations adopting ISO 14001, companies that fail to follow the lead may lack access into such markets and even when granted access, may not be able to compete in an environmentally conscious market.

We end our discussion in this chapter by presenting two classic case studies from two of the world's leading companies in sustainable manufacturing namely Kodak and Xerox. There are a host of other companies that have very good environmental strategies but the success stories of Kodak and Xerox may provide a motivation for more companies to reevaluate their environmental programs and perhaps, benchmark the leaders.

## References

Giuntini, R., "Redesign it + Produce it + Rent it + Support it + Renew it + Reuse it = Reverse Logistics Reinventing the Manufacturer's Business Model" APICS—The Educational Society for Resource Management, 1997 (http://www.apics.org/SIGs/Articles/redesign.htm).

"Kodak Recycles Its 50 Millionth FUN SAVER" http://www.kodak.com/US/en/corp/environment/performance/recycling/suc.shtml, retrieved 3/2/99.

Furukawa, Y., "'Throwaway' mentality should be junked," *The Daily Yomiuri*, October 1, 1996, pp. 10.

Fukukawa, S., "Japan's policy of sustainable development," *Columbia Journal of World Business*, 27 (3 & 4), 96-105.

Umeda, Y., "Research topics of the Inverse Manufacturing Laboratory," http://www.inverse.t.u-tokyo.ac.jp, 1995.

Madu, C.N., "A decision support framework for environmental planning in developing countries," *Journal of Environmental Planning and Management*, Vol. 43 (3), 287-313, 1999.

Madu, C.N., Managing Green Technologies for Global Competitiveness, Westport, CT: Quorum Books, 1996.

Madu, C.N., "Introduction to ISO and ISO quality standards," in Handbook of Total Quality Management (ed., C.N. Madu), Boston, MA: Kluwer Academic Publishers, 1999, pp. 365-387.

Brundtland Commission, Our Common Future, Geneva, Switzerland: Report of UN World Commission on Environment and Development," April, 1987, p. 43.

Yoshikama, H., "Sustainable development in the 21st century," http://www.zeri.org/texter/ZERT_96_industries.html, 1996.

Singer, S.F., "Sustainable development vs. global environment – Resolving the conflict," *Columbia Journal of World Business*, 27 (3 & 4), 155-162.

Schmidheiny, S., "The business logic of sustainable development," *Columbia Journal of World Business*, 27 (3 & 4), 18-24.

Roy, R., and Whelan, R.C., "Successful recycling through value-chain collaboration," *Long Range Planning*, 25 (4), 62-71, 1992.

Dillion, P., and Baram, M.S., "Forces shaping the development and use of product stewardship in the private sector," Conference on the Greening of Industry, The Netherlands, 1991.

Duncan, N.E., "The energy dimension of sustainable development," *Columbia Journal of World Business*, 27 (3 & 4), 164-173.

"UsingEco-labelling," http://www.uia.org/uiademo/str/v0923.html, retrieved 2/6/99.

Henriksen, A., and Bach, C.F., "Voluntary environmental labeling and the World Trade Organization," International Trade, Environment and Development, Institute of Economics, Copenhagen University, November 9, 1998.

Christensen, J.S., "Aspects of eco-labeling on less developed countries," International Trade, Environment and Development, Institute of Economics, Copenhagen University, December 7, 1998.

Affisco, J.F., "TQEM—methods for continuous environmental improvement," in Handbook of Total Quality Management (ed., C.N. Madu), Boston, MA: Kluwer Academic Publishers, 1999, pp. 388-408.

"Sustainable product development—The First 30 years," http://www.xerox.com/ehs/1997/sustain.htm, page 8

"What is the WBCSD," http://www.wbcsd.ch/whatis.htm, retrieved 3/9/99.

"Green business networks," http://www.epe.be/epe/sourcebook/1.14.html, retrieved 3/9/99.

Robinson, A., "Inverse manufacturing," March 10, 1998, http://mansci2.uwaterloo.ca/~msci723/inverse_mfg.htm

# Chapter 3

# Environmentally Conscious Manufacturing

In Chapter 2, we introduced the concept of sustainable development and sustainable manufacturing. It is apparent that the goals of sustainable growth cannot be attained if we do not change our consumption pattern. Manufacturing plays a critical role in introducing new products to the market and also in shaping our tastes and consumption. Sustainable development cannot be achieved if adequate emphasis is not paid to sustainable manufacturing.

Environmentally conscious manufacturing (ECM) is often referred to as "ecofactory." The goal of ecofactory is to achieve optimal utilization of natural resources without harming the environment and without compromising the quality of the products. This goal can be achieved if effective utilization of natural resources is made, waste is minimized, and cradle-to-grave approach of the product is taken. In other words, the goal of ecofactory is not limited to the production process but spans through distribution, consumption and recovery and effective disposal of potential wastes. Thus, the manufacturer tracks and manages the whole product life cycle. Environmentally conscious manufacturing (ECM) as a strategy adopts a systemic approach to product development and distribution. This approach starts from the design stage of the product where every effort is made to ensure that the product is environmentally friendly to an environmentally responsible disposal of the product or its waste at the end of the product life cycle. Watkins and Granoff (1992), define environmentally conscious manufacturing as "those processes that reduce the harmful environmental impacts of manufacturing, including minimization of hazardous waste, reduction of energy consumption, improvement of materials utilization efficiency, and enhancement of operational safety." Through ECM, the aim is to achieve "zero waste" through total system integration of the entire production and distribution processes. A preventive approach to environmental protection that focuses on reducing waste at the source rather than at the end-of-the-pipe treatment is commonly adopted. Manufacturers develop products for ease of disassemble, recycling, and use non-hazardous and non-toxic materials. Some of the strategies for environmentally conscious manufacturing such as inverse manufacturing, recycling, reverse logistics, re-manufacturing, and others, were briefly introduced in chapter two. In this chapter, we explore in greater depth some of the most commonly used strategies.

Reverse Logistics—also known, as reverse supply chain management has increasingly become popular among manufacturers. Have you ever wondered what manufacturers do with products that are returned by retailers? Well, this is the case of reverse logistics. Reverse logistics deals with the processing of goods that are returned from the customer's customer. The normal process for supply chain management involve the flow of goods and services to the consumer with little or no focus on the flow of waste back to the manufacturer. This is changing hence the term reverse supply chain management. The increasing cost of landfills, environmental laws and regulation and the economic viability of environmental strategies are pushing manufacturers to now consider reverse supply chain management. For example, localities are establishing new landfill regulations that often require separation and grouping of materials of the same type for ease of recycling. Recyclable items are no longer grouped together with all types of waste and garbage. Also communities are developing designated drop sites and manufacturers develop disposition processes. Some have also developed recycling programs for containers and cans with designated sites for ease of management of the recycled items. Here, the customer through a reverse chain returns goods to the manufacturer for effective disposal. For reverse logistics to work effectively, information management is critical. Manufacturers embarking on reverse logistics must be able to sort out salvageable items and separate repairable and non-repairable salvageable inventories. Effective management planning system that focuses on transportation planning, location analysis, and inventory control and management, and coordination of customer and vendor activities will be needed. Thus, it is important that in reverse logistics, not only does the product flow back to the manufacturer but also, information about the good that is being returned should flow back to the manufacturer. Manufacturers may be able to improve their bottom line if reverse logistics strategy is successfully implemented. However, before we continue, an important question in the context of environmental planning and management is why is reverse logistics an acceptable alternative to environmental protection?

Marien[1999], notes that manufacturers are developing source-reduction strategies as promising alternative to minimizing wastes and environmental pollution. This strategy is based principally on:

- Reducing the weight and size of the product. This optimizes the logistics costs in both the supply chain and the reverse supply chain. Further, the cost of warehouse space is reduced as the size is reduced. Also, labor and material handling costs are significantly reduced when the item is trimmed in size and weight. Many organizations are embarking on this strategy. For example, the packaging industry is increasingly, achieving reduction in their packaging program. Sears for example, has reported a packaging reduction program that has saved 1.5 million tops in the supply chain which is a savings of about $5 million annually in procurement and disposal costs [Marien, 1999]. Likewise, computer companies are increasingly building faster and more effective computers that are smaller in size and weigh less.

- Minimization of production and distribution operations. There are many ways this could be achieved. First, minimizing production operation can be achieved by designing and building the right products that are highly dependable for the consumers. The high quality built into the product means that there will be less rejects, reworks, or returns. Thus, limited resources are optimally utilized and energy consumption is reduced. Further, by doing things right the first time, labor cost is reduced. With the high quality of the product, it becomes competitive and the organization gains. Distribution operation is also optimized when quality is built into the product. Clearly, the high return rate of products will be avoided thus reducing the high cost of distribution through the supply chain. Also, there will be less need for inventory of replacement parts and returned goods and more efficient use of the distribution channel. The end result is that resources are optimally utilized and waste is minimized.

- Reuse of materials and resources. The Eastman Kodak single-use-camera is the best example of this concept. However, there are many more examples. Computer components are easily recycled and many of the paper products used today are recycled. It is possible to reuse some of these materials as in the case of Kodak and Xerox because they are designed for ease of disassembly. Thus, when a product is returned, it is easy to disassemble it, recover usable parts and integrate them in the production process. The concept of re-manufacturing is getting popular today because it is easier to recover useable materials from used equipment.

- Another strategy is the substitution of materials that are environmentally friendly. This strategy is mostly utilized when a hazardous or toxic material is replaced with a more environmentally friendly substance. It could also be applied in conserving resources that are very limited in supply. One example is the replacement of the use of DDT (dichloro-diphenyl-trichloro-ethane) as a pesticide. DDT is a chlorinated hydrocarbon and it is not easily biodegradable. When used, it can be found in the tissues of living organisms that are exposed to it. It also has a disastrous influence on marine life as it reduces the rate of photosynthesis in marine phytoplankton, which is the base for most marine food chains. Since humans are at the end of this food chain, they can suffer irreparable health conditions from deposits of DDT in their tissues. Another example is the worldwide ban of the use of CFCs, carbon tetrachloride, and methyl chloroform. The U.S. Clean Air Act of 1990, outlawed these chemicals in the year 2002. CFCs are normally used as coolants and were once common in home refrigerators but are also ozone depleting. Manufacturers such as DuPont have already replaced CFCs with hydrofluorocarbon (HFC) called HFC-134a. This new product is nonflammable, non-toxic, and non-ozone depleting, and has the same energy efficiency as CFCs.

## Why Reverse Logistics?

There is limited landfill space available for dumping of wastes. Also, landfill is becoming increasingly more expensive to manage. Many organizations are realizing that reverse logistics offers the opportunity to recycle and reuse product components while cutting down the cost and the amount of waste that will normally be incurred. In Chapter 2, we presented a case of Kodak single use camera, which follows a reverse logistic strategy. We noted that in the U.S., a 63 percent return rate has been achieved for recycling. This is equivalent to fifty million SUCs or enough SUCs to fill up 549 tractor-trailer loads. Imagine the landfill requirement for disposing of waste of such an enormous quantity. Not only was waste avoided, Eastman Kodak improved its bottom-line by recycling and reusing components from returned SUCs and also, reduced the cycle time for re-introducing the product into the market. While the strategy is environmentally responsible, it is also economically profitable. Eastman Kodak is not alone. Many other organizations are adopting reverse logistics. Hewlett-Packard for example refill returned printer toner cartridges and Xerox recovers used machines from customers and use them to remanufacture new ones. These actions have reduced the demand for landfill, reduced the need for excavation of new raw materials, and have reduced energy consumption from the processing and manufacturing of virgin products.

New environmental laws and regulations are clear in assigning responsibilities to manufacturers. Manufacturers must now take full responsibility of their products through the product's life cycle, or they may be subject to legal action. For example, new laws regarding the disposal of motor or engine oil, vehicle batteries and tires assign disposal responsibility to the manufacturer once these products have passed their useful life. Thus, as Marien [1999] notes, to avoid the related supply chain complexity, it is important for manufacturers to build reusability into their products. Thus, manufacturers act by developing infrastructure to handle post-distribution and consumption activities. Hence, reverse logistics is increasingly seen as a competitive strategy that is not only designed to meet the social responsibility function of the firm but also designed to make the organization more profitable. Marien [1999] points out that the savings accruing to organizations that adopt reverse logistics are in the form of savings from "raw material and packaging procurement, manufacturing, waste disposal, and current and future regulatory compliance."

Businesses look at their bottom-line. Ultimately, the goal of environmental protection cannot succeed without the participation of business organizations. For their cooperation and participation in environmental programs to be assured, there must be potential benefits to them. In the past, businesses use to view environmental protection efforts as wasteful expenditure but not any more. They are now seeing that environmental programs offer competitive advantage. More consumers pay attention to the environmental friendliness of the products they purchase. Also, organizations are beginning to realize that environmental strategies such as reverse logistics can cut down drastically on production and operations cost thus improving their profit margins. Some of the costs incurred from reverse logistics include the costs of refuse, reworks, recyclables, rejects, reprocessed overruns, reuse, remake, redo, residues, reorder, resale, returnable shipping containers and pallets. However, some of these costs are controllable. For example, the cost of rejects, reworks, and reprocessing can be avoided if the organization adopts a quality imperative. Thus, reverse logistics operates efficiently when the organization adopts other comparative strategies such as developing an effective quality program. Also, the benefits of reproducing a product from recyclable items may far exceed the costs associated with reverse logistics. Some of these costs such as the cost of disassembles could be seen as production costs since they replace the traditional costs of production. However, the organization can become more effective by designing its products for ease of assemble and also, by developing an efficient reverse supply chain network.

There are several logistical problems involved with reverse supply chain network. For example, what is the cost of transporting the goods back to the manufacturer? How often can the goods be returned? Is it better to use decentralized or centralized reverse logistics strategy? What is the cost of inventory? And what is the cost of processing the returns? To address some of these issues, Bunn [1999] presents factors for consideration in developing centralized logistics strategy. These factors focused mainly on costs relating to store labor processing, transportation, inventory, opportunity costs, credit terms, and operating cost of a centralized facility. These factors may come into play in negotiating better terms with vendors. By using the right logistics strategy, costs can be significantly reduced. However, each operation is different and it is important to take its uniqueness into perspective in determining the correct reverse logistics strategy.

## Environmental Action Box

We shall present a success story on the use of reverse logistics. Our example here is the giant name-brand cosmetics manufacturer Estee Lauder Companies Inc.

*Estee Lauder Companies Inc.:*

Estee Lauder Companies Inc. was founded in New York City in 1946 and is one of the world's leading manufacturers and marketers of cosmetics. Among its popular brands are Estee Lauder, Clinique, Prescriptives, Aramis, Tommy Hilfiger, Origins, and Jane. The company estimates that about $60 million worth of returned products from retailers were being dumped in landfills each year [Caldwell, 1999]. This amounts to about a third of returned goods from retailers. To cut down on this tremendous waste, the company decided to embark on developing processes and information technology. Reverse logistics became a viable alternative to consider. To develop the reverse logistics system, Estee Lauder Companies Inc. invested $1.3 million on scanners, business-intelligence tools and an Oracle data warehouse. Estee Lauder also customized the software to process returned goods. The results were encouraging. Some of the benefits derived by the company through the application of reverse logistics are outlined below [Caldwell, 1999]:

- The company was able to evaluate 24% more returned products, redistribute 150% more of its returns, and saved $475,000 annually in labor costs.

- The number of returned products destroyed because they exceeded their shelf life dropped from 37% to 27% in 1998. It is expected that this number will eventually drop to 15%.
- The processing time for returned products was significantly reduced and the time to introduce returned goods to the market became faster.
- Production and inventory levels reduced.
- Better product management information system existed as data on the reasons for returns were collected.
- Information system in place helped in developing better marketing, packaging, and production strategies.

Reverse logistics is an effective environmental management strategy. Its role in environmental protections can be outlined as follows:

- It helps to better manage and conserve landfills as returned goods are disassembled, reused and reintroduced into the market place.
- Information collected through the process can help guide the production process, quality control programs, and marketing strategies. This will help ensure more efficient use of resources.
- Energy consumption is reduced, as waste management becomes more effective.
- It complements the efforts of other environmental programs such as recycling by ensuring that valuable materials and resources are reused in the production process.
- Reverse logistics is a win-win strategy where the organization stands to gain by making wastes profitable and the society in general stands to gain as waste disposal is minimized.

## Recycling

Recycling is a process of converting materials that could have been treated as wastes into valuable resources. There are many examples of recycling such as aluminum cans, bimetal cans, glass bottles, newspapers, paper products, and composting. Recycling is one of the better-known strategies for environmentally conscious manufacturing. In fact, the concept of recycling is vogue today as many communities have adopted recycling programs. In these areas, recyclable materials are carefully separated from ordinary garbage or waste and the garbage collectors make a distinction between recyclable materials and garbage when they schedule pickups. Recycling is considered an environmental success story of the 20$^{th}$ century. As the United States Environmental Protection Agency (EPA) reports, recycling including composting has contributed to a significant reduction in the amount of material being turned over to landfills and incinerators for disposal [downloaded 11/8/99, pp. 1-4]. Based on this account, in 1996, 57 million tons of material that would have been sent to landfills and incinerators as garbage were recycled. This amounts to a 67% increase from the 34 million tons that were recycled in 1990. Likewise, the number of curbside collection programs in the United States has increased dramatically. There are reasons for the success of recycling:

- Consumers are increasingly concerned about the depletion of earth's limited resources and are also worried about the degradation of earth's environment through landfills, excavations, destruction of forestry, and pollution of air, water and land. They are therefore, willing to participate in protecting the environment. It is their cooperation that has attributed to the great successes of recycling programs. Consumers are now buying recycled products and investing in companies that market recyclable products.

- Recycling is profitable. Many organizations are now realizing that they can cut down on cost of material, reduce the cycle time for introducing new products, reduce processing time, and even become more efficient in their planning process if they recycle and reuse their products. They have better control of their recycled products and may avoid complex supply chain networks associated with dealing with vendors for virgin products who are    outside their control.

- Environmental laws and regulations that require that certain types of products be recycled have also contributed since the penalty for non-compliance may at times be severe. Apart from the legal ramifications of non-compliance, environmental activist groups can also damage the reputation of non-complying companies thus contributing to high customer dissatisfaction with the company and its products and services.

- Apart from the profitability of recycling programs to organizations, consumers benefit directly. For example, another form of recycling is composting. Composting is the recycling of organic wastes. Many of the organic wastes can be easily recycled such as food and yard wastes and can be fed back to soils or applied in landscapes. Such applications help reduce plant diseases and provide nutrients to soil. Further, some of the beneficial soil organism such as worms and centipedes feed on such wastes.

Palmer [2000] gave a detailed discussion and definition of recycling and also identified the conflicts in current recycling programs. Noting that we live in a world endowed with finite resources, it is important to articulate and develop resource policy that can help achieve sustainable development while protecting the environment. Although this is often difficult to achieve due to several pressures on national economic programs that for some countries, often demand exploitation of these limited natural resources to generate needed capital however, it is imperative that national planning issues focus also on the needs of the future generation. Such focus will help to seek for example better alternatives to the use of landfills and encourage recycling policies that are environmentally friendly. For example, prior to recycling, all "wastes" were grouped as the same and are dumped in designated dumping sites for wastes. When a landfill has been used up, a new one is created and this process goes on and on. Little did the general public know that apart from the unsightly image of the landfill and the unbearable odor gasping out from it, it could also become a health hazard. Forty percent of the Superfund sites are municipal garbage dumps [Palmer, 2000], since all sites were for management of toxic wastes, disposal of chemicals and other hazardous wastes. For example, approximately one-third of GM 's toxic release inventory to landfill from foundry waste used to contain zinc. A new plan by GM will eliminate these land releases from GM foundries by the year 2002 [Annual report, 1997]. In 1997, GM recycled 61 percent of all these wastes. Furthermore, the separation of wastes into "recyclables" and "non-recyclables" has contributed immensely to sustainability. First, there is lesser need for landfills since the amount of wastes designated for dumpsites have declined. Second, recyclable items have extended lives and are re-used in the manufacturing and production processes. This use decreases the need for new or virgin products similar to the recycled item. Third, there is less need for energy consumption. As treated equally then, toxic compounds and chemicals were equally mixed with other wastes. This has been attributed to many of the environmental degradation, destruction of wildlife, and health problems. Thus, it became clear that these "wastes" needed to be separated especially from their sources.

Although recycling can help conserve resources and save energy, not all materials are easily recycled. For example, cadmium and beryllium are not easy to recycle. Cadmium is widely known to the general public for its use in batteries. Its application in nickel-cadmium (Ni-Cd) batteries is one of the easiest forms to recycle. Many other applications of cadmium are in low concentrations and are difficult to recycle since much of the cadmium is dissipated. However, the growing application of cadmium in batteries and the concern about potential environmental pollution has led to regulations limiting the dissipation of cadmium into the ground [1997]. Moreover, Beryllium is also difficult to recycle and it is widely dispersed in products when it is used in manufacturing. It also dissipates and is very difficult to recycle.

## Recycling Statistics

Recycling is a worldwide phenomenon. People all over the world are paying attention to recycling. Available statistics in the industrialized countries tend to support the growth in recycling programs. U.S. Geological Survey [1997] presents recycling statistics for selected metals. Generally, the survey tends to support the growth in recycling efforts. Aluminum scrap is one of the popularly recycled materials since aluminum is widely used in the manufacture of beverage cans. Thus, used beverage can (UBC) is one of the major sources of aluminum scrap. Data obtained for the survey from the Aluminum Association Inc., the Can Manufacturers Institute, and the Institute for Scrap Recycling Industries suggests that 66.8 billion of aluminum were recycled in the United States in 1997. A 66.5% recycling rate that is based on the number of cans shipped during the year was obtained. This is an increase from the 63.5% recycling rate obtained in 1996. Further, domestically produced aluminum beverage cans in 1997 had an average 54.7% postconsumer recycled content. According to the Aluminum Association Inc., [1998], this is the highest recycled content percentage of all packaging materials. Statistics on other metals that were recycled between 1993 and 1997 are also presented in this survey [1997].

Aluminum cans are widely used as soft drink containers, and are perhaps, one of the oldest forms of recycling. The National Soft Drink Association presents some startling statistics on the recycling of soft drink containers from 1990–1997. They present the following data [1999]:

- Soft drink container recycling has risen from 48.7% to about 60% since 1989.
- In 1997, a total of 51.9 billion soft drink containers were recycled.
- Soft drink containers comprise of less than 1 percent of solid waste disposed in the U.S.
- Beverage containers account for less than 20 percent of the materials collected through curbside recycling programs but generate about 70 percent of total scrap revenue.
- Through packaging innovation, the weight of soft drink containers has been reduced by about 30 percent since 1972.
- 22 percent of soft drinks are dispensed from fountains while the balance is packaged.

These data show remarkable improvement in the recycling of soft drink cans partly through cooperative recycling programs and product packaging innovations. We have also analyzed graphically, the statistics provided by the National Soft Drink Association (NSDA). This data is presented below as Figure 3.1.

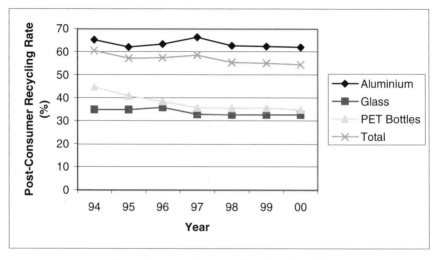

*Figure 3.1: Post-Consumer Recycling Rate (%)*

## 1991-2000 Soft Drink Container Recycling Figures

Prepared by NSDA Environmental Affairs Department

### Total Units Shipped (billion units)

| Container Type | '94 | '95 | '96 | '97 | '98 | '99 | '00 |
|---|---|---|---|---|---|---|---|
| Aluminum Cans[1] | 61.2 | 64.6 | 64.3 | 66.1 | 69.6 | 68.9 | 67.8 |
| Glass Bottles[2] | 3.6 | 2.2 | 1.0 | 1.0 | 0.8 | 0.8 | 0.8 |
| PET Bottles[3] | 12.8 | 16.0 | 18.7 | 21.3 | 23.7 | 25.2 | 25.6 |
| TOTAL | 77.6 | 82.8 | 84.0 | 88.4 | 94.1 | 94.9 | 94.2 |

### Total Units Recycled (billion units)

| Container Type | '94 | '95 | '96 | '97 | '98 | '99 | '00 |
|---|---|---|---|---|---|---|---|
| Aluminum Cans[1] | 40.0 | 40.2 | 40.8 | 44.0 | 43.6 | 43.1 | 42.1 |
| Glass Bottles[2] | 1.3 | 0.8 | 0.4 | 0.3 | 0.3 | 0.3 | 0.3 |
| PET Bottles[3] | 5.7 | 6.6 | 7.2 | 7.6 | 8.4 | 9.0 | 9.0 |
| TOTAL | 47.0 | 47.5 | 48.4 | 51.9 | 52.3 | 52.4 | 51.4 |

## Post-Consumer Recycling Rate (%)

| Container Type | '94 | '95 | '96 | '97 | '98 | '99 | '00 |
|---|---|---|---|---|---|---|---|
| Aluminum Cans[1] | 65.4 | 62.2 | 63.5 | 66.5 | 62.8 | 62.5 | 62.1 |
| Glass Bottles[2] | 35.0 | 35.0 | 36.0 | 33.0 | 32.8 | 32.8 | 32.8 |
| PET Bottles[3] | 44.9 | 41.0 | 38.6 | 35.8 | 35.6 | 35.7 | 35.0 |
| TOTAL | 60.6 | 57.4 | 57.6 | 58.7 | 55.6 | 55.2 | 54.6 |

<u>Sources:</u> (1) **aluminum data** – Can Manufacturers Institute, Aluminum Association, Institute of Scrap Recycling Industries, Steel Recycling Institute; (2) **glass data** – Glass Packaging Institute, U.S. Department of Commerce, Current Industrial Reports; (3) **PET data** – American Plastics Council, Container Consulting, Inc. (*Reprinted with permission from NSDA Environmental Affairs Department.*)

As Figure 3.1 shows, even though there seems to be some gains in post-consumer recycling rate from 1996 to 1997, however, this rate is unstable over the time period considered. For example, there was a slight drop between 1994 and 1995 and then a dramatic increase to 1997 and a drop again in 1998. This up and downward swing appears to be prominent with aluminum cans. Post-consumer recycling rate for glass bottles seems to be relatively more stable. While generally, there are some gains made in improving post-consumer recycling rate, however, continuous improvement is needed to avoid the type of instability noted in the data. In addition, new studies need to identify the causes of such swings.

### Environmental Action Box

Paper recycling is one of the most popular forms of recycling. In the Environmental Action Box, we use the giant paper manufacturer – International Paper as an example.

### International Paper (IP)

International Paper is a leading manufacturer of paper. It is actively engaged in paper recycling programs with nearly $700 million invested. Further, the quality of its recycled paper is so high that it is indistinguishable from non-recycled paper. However, the major challenge facing IP and other paper manufacturers is that the quality of post-consumer recycled paper gradually deteriorates to the point that the recycled fibers become useless. This will eventually require virgin fibers to generate new paper products. Thus, if alternative to the use of wood pulp for paper manufacturing is not developed, sustainability cannot be achieved in the long run. International Paper has opened up the discussion and suggestion on alternatives to this problem to the general public. One alternative that is frequently alluded to is the use of annual fibers to make good paper. Although this sounds plausible, it has limitations and serious constraints that need to be considered. Some of the problems are:

It could lead to more environmental degradation. For example, farmers that currently manage forests because of high timber values may be tempted to mow down the timbers and harvest annual fibers if demand and price for timber drop down significantly. Thus, any alternative offered as substitute to the use of timber should recognize the economic hardship forest managers may face and a plan must be in place to accommodate them.

Annual fibers need to be grown in large quantities to meet the current demand for paper. This will require land and perhaps, displacement of other harvested items or wildlife, which again can potentially affect the ecobalance.

Decisions to substitute annual fiber to timber have economic ramifications. The paper industry is one of the most capital intensive. Such decisions will require reinvestment in new technologies, equipment and material handling processes that may be different from what the industry is currently used to.

We shall now discuss some of the other environmentally conscious sustainable manufacturing practices.

## Inverse Manufacturing

Inverse manufacturing has its roots from Japan where it began as a reuse and recycle project. The concept of inverse manufacturing is an extension of the recycling, reuse, and remanufacturing concept. It focuses on the pre-manufacturing process especially at the product design stage. The aim is to prolong the useful life of the product through design by designing reuse and recycling features into the product. The other feature is to design the product so that it is easy to disassemble. One way this is accomplished is by building modules into a product. For example, computers and refrigerators are made up of modules. These modules can be upgraded or replaced without replacing the entire product. For example, the functions of a personal computer can be upgraded by replacing modules such as the central processing unit (CPU) [1996]. In addition to these attributes, inverse manufacturing focuses a lot on maintenance. It envisages leaner manufacturing where companies will have to do away with the concept of mass production by creating quality products that will last longer. This vision will require transformation of many of the manufacturing outfits into life cycle companies with a focus on providing maintenance services on their products. The construction industry is actually an industry that survives well by providing mostly maintenance services on existing infrastructures. Through inverse manufacturing, product manufacturers can in fact, transform themselves to life cycle companies by providing maintenance operations and services to their products, thereby prolonging the useful life of the product. Why this concept may seem radical, it may be a desired option given the increasing problem with landfills and the limited natural resources. This closed-loop product life cycle approach leads to minimal disposal and environmental costs. According to the Inverse Manufacturing Forum Secretariat, inverse manufacturing takes a reverse process approach by focusing on the recovery of the product to disassembling to reutilization and production. This gives a complete loop of the product life cycle.

The concept of inverse manufacturing also requires a cultural transformation. The general society must be educated on the need to maintain products rather than discarding or dumping them in landfills.

Further, manufacturers should also educate their customers and support the initiative to prolong the lives of these products. One factor that worries some about inverse manufacturing is that the decline in mass production may lead to loss of job as production capacity is decreased. However, the transformation to life cycle industry may absolve the excess capacity that may result from the decline in production.

In sum, inverse manufacturing involves the following:

1. Integration of reuse and recycling plans at the early stages of product design,
2. Emphasis on product maintenance and reduction in production volume through transformation to life cycle industry, and
3. Modular design strategies to make it possible to expand and upgrade product functions. Hata [1997] presents a good framework on inverse manufacturing. An adapted version is presented below:

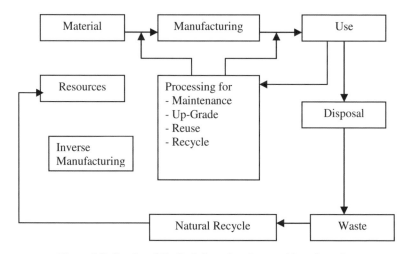

*Figure 3.2: Product Life Cycle based on Inverse Manufacturing*

### Remanufacturing

Remanufacturing is the process of rebuilding a product from ground up utilizing new parts to return it to a condition "as good as new." This involves disassembly of each component in a product, inspection and testing of parts, evaluation of components for quality and reliability standards, replacement and upgrade of parts. Remanufactured parts possess some important attributes such as

- The appearance of the product is refurbished and enhanced
- Quality and reliability standards are satisfied
- Performance is guaranteed
- They are economical and environmentally friendly.

Remanufacturing is different from refurbishing where many of the changes are cosmetic involving rehabilitation of an older product to provide services, retrofits and upgrades. Refurbishing does not undergo the extensive process of remanufacturing a product to ensure that it is returned to a condition that is "as good as new."

## Design for Environment

The strategies discussed above consequently lead to the design for the environment. A combination of strategies must be followed in order to achieve an improvement in environmental performance. These strategies include raw material acquisition, product design, product usage and disposal. At each stage of the product life cycle, attention must be paid on how to conserve and effectively use resources. Pollution prevention and waste minimization become the driving force of strategies for designing for the environment. The product design stage pays tremendous attention on how limited resources are used, explores the use of replenishable materials and the use of substitute products to prevent pollution problems, and considers the energy demands of the product. The product design stage is critical in environmental management since problems uncovered at this stage can spread through the product life cycle and create more environmental hazards. It is at the product design stage that the decision to design for recyclability is made. If the product is not effectively designed to be easily disassemble and for component recovery or for material separability to avoid contamination, it will be difficult to treat this problem later on. Thus, a strategy at this stage to design for recyclability can help in minimizing both material and energy wastes and ensuring the conservation of limited resources. This strategy of designing for recyclability is also closely linked to the design for remanufacture. When components can be easily recovered from a malfunctioning unit, they could also be equally remanufactured, restored and reused. The extension of a component's life cycle also implies that there will be less demand for landfills for disposal of old units and there will also be less need for virgin products. These strategies therefore supplement each other. When a product is effectively designed for recyclability, it will meet the remanufacturing needs and its disposal needs.

## Conclusion

This chapter discussed some of the strategies that are currently being adopted to achieve environmentally conscious manufacturing. It discusses how companies are designing for environment by using environmentally sound principles such as recycling, remanufacturing, inverse manufacturing and reverse logistics. It shows that these strategies are not only environmentally sound but also increasingly profitable to corporations. Recycling statistics from the National Soft Drink Association also showed that recycling is increasingly a worldwide phenomenon as more and more corporations are adopting a cradle-to-grave approach to their products. This chapter also presents environmental action boxes featuring environmental activities at Estee Lauder Companies Inc. and International Paper.

## References

Watkins, R.D., and Granoff, B., "Introduction to environmentally conscious manufacturing," *International Journal of Environmentally Conscious Manufacturing*, 1 (1), 5-11, 1992.

Bunn, J., "Centralizing reverse logistics: How to understand if it will work for you," downloaded 11/9/99, http://usserve.us.kpmg.com/cm/article-archives/actual-articles/revlogis.htm, pp. 1-3.

Marien, E., "Reverse logistics as competitive strategy," downloaded 11/9/99. http://www.apics.org/SIGs/articles/marien.htm, pp. 1-10.

Caldwell, B., "Reverse Logistics Untapped opportunities exist in returned products, a side of logistics few businesses have thought about--until now," Information Week Online, http://www.informationweek.com/729/logistics.htm, pp. 1-4, 4/12/1999.

"Produce less waste by practicing the 3Rs," http://www.epa.gov/epaoswer/non-hw/muncpl/reduce.htm#recycle, downloaded 11/8/99, pp. 1-4.

Palmer, P., "Recycling as universal resource policy" in Madu, C.N., Handbook of Environmentally Conscious Manufacturing, Boston, MA: Kluwer-Academic Publishers, pp. 205-228, 2000.

GM Annual Report, 1997

U.S. Geological Survey -- Minerals Information -- 1997, "Recycling – Metals," pp. 1-13.

Aluminum Association Inc., "Aluminum can recycling rate reaches 66.5 percent: Washington, DC, *Aluminum Association News*, March 6, 1998, pp. 1-4.

National Soft Drink Association, "Recycling," http://www.nsda.org/Recycling/facts.html, 1999, p.1.

Yoshikama, H., "Sustainable manufacturing in the 21st century" From ZERI (Zero Emissions Research Initiatives) Symposium, May 1996, http://www.zeri.org/texter/ZERT_96_industries.html.

Hata,Tomoyuki, Kimura, Fumihiko, and Hiromasa Suzuki (1997) "Product Life Cycle Design based on Deterioration Simulation," www.cim.pe.u-tokyo.ac.jp/~lcgroup/theses.htm downloaded March 2, 1999.

**Chapter 4**

# The ISO 14000 Model

In this chapter, we shall discuss some of the most important standards for environmental management systems. Such standards are embodied in what is now known as ISO 14000 family of standards. These are set of standards and guidelines that could help businesses to develop more environmentally friendly products and services. ISO standards have received worldwide attention primarily because of the reputation of ISO (International Organization for Standardization) itself. The origins of ISO dates back to 1947 when it was formed as an NGO (non-governmental organization) with the purpose of promoting the development of standards to facilitate the international exchange of goods and services. ISO seeks international cooperation in scientific, technological and economic activities. Its membership has grown to over 100 countries that are represented by their national standards organization. The term ISO is derived from the Greek word 'isos' which means 'equal.' This can explain the goal of ISO to develop "equal" standards to guide the international exchange of goods and services. International standardization of goods and services protects the consumer and may also facilitate the transfer of technology and trade. Some of the benefits are in (Introduction to ISO from ISO Online):

-   Enhanced product quality and reliability at reasonable price;
-   Improved health, safety and environmental protection and reduction of waste;
-   Greater compatibility and interoperability of goods and services;
-   Simplification for improved usability;
-   Reduction in the number of models and thus reduction in costs;
-   Increased distribution efficiency and ease of maintenance.

In today's global economy, there is a need for standardization both in product quality and environmental content. With uniformity in standards among similar industries and technologies, companies can compete on a level playing field by removing some of the technical barriers to trade. However, achieving some of the standards may in the short-run become very costly and may make it difficult for some poorer nations to participate effectively in global markets.

The ISO successfully developed the international standards on quality assurance techniques and practices in the 1980s. These standards known, as ISO 9000 series of standards for product quality got worldwide acclaim and has fueled the development of a new set of standards for environmental management systems.

## ISO 14000 Series

The ISO 14000 series of standards represent new sets of standards on environmental quality issues. They deal with guidelines and principles of environmental management systems to make businesses to focus on the growing need of environmental protection. The concept of ISO 14000 was introduced by a team of 50 business executives interested in sustainable development and known as the Business Charter for Sustainable Development (BCSD). By 1992, the world was increasingly concerned about the increasing pollution of the natural environment. The Earth Summit conference on Environment and Development was organized by the United Nations and held in Rio de Janeiro, Brazil in response to these concerns. ISO then formed the Strategic Advisory Group on the Environment (SAGE) and charged it with the evaluation of the international standards on environmental management systems. SAGE's recommendations in 1993 led to ISO 14000. Technical committee (TC) 207 was then formed to replace SAGE. This committee has the responsibility to develop standards for global environmental management systems and tool. The committee was to focus on the following areas of environmental management systems:

- Environmental management systems (EMS);
- Environmental auditing;
- Environmental labeling;
- Environmental performance evaluation (EPE);
- Life cycle assessment;
- Terms and definitions;
- Environmental aspects in product standards (EAPS).

By the third quarter of 1996, the committee completed its work and published a series of standards to help firms manage and evaluate the environmental aspects of their operations. In Tables 4.1, we present the ISO 14000 family of standards and their applications. This table is adopted from the ISO web site (http://www.iso.org).

*Table 4.1: ISO 14000 Series Standards*

| Standard number | Title |
|---|---|
| ISO 14000 | Environmental management systems — general guidelines on principles, systems and supporting techniques |
| ISO 14001 | Environmental management systems — specifications with guidance for use |
| ISO 14004 | Environmental management systems — general guidelines on principles, systems and supporting techniques |
| ISO 14010 | Guidelines for environmental auditing — general principles of environmental auditing |
| ISO 14011 | Guidelines for environmental auditing — audit procedures — part 1: auditing of environmental management systems |
| ISO 14012 | Guidelines for environmental auditing — qualification criteria for environmental auditors |
| ISO 14020 | General principles for all environmental labels and declarations |
| ISO 14021 | Environmental labels and declarations — self-declaration environmental claims — terms and definitions |
| ISO 14022 | Environmental labels and declarations — self-declaration environmental claims — symbols |
| ISO 14023 | Environmental labels and declarations — self-declaration environmental claims — testing and verification |
| ISO 14024 | Environmental labels and declarations — self-declaration environmental claims — type I guiding principles and procedures |
| ISO 14031 | Environmental management — environmental performance evaluation guideline |
| ISO 14040 | Life cycle assessment — principles and framework |
| ISO 14041 | Life cycle assessment — inventory analysis |
| ISO 14042 | Life cycle assessment — impact assessment |
| ISO 14043 | Life cycle assessment — interpretation |
| ISO 14050 | Terms and definitions |
| ISO 14060 | Guide for the inclusion of environmental aspects in product standards |

There have been significant changes in the ISO 14000 series of standards since 1996. Updated lists at different stages of development are presented in Tables 4.2 and 4.3 below.

The process for adopting a standard is briefly discussed so that the reader can see from the prefixes attached in the tables below, the status of a particular working document.

Before a committee's draft is accepted as a standard, it must be approved following a consensus process. Briefly, the following steps are taken:

- A working draft (WD) is developed by a work group (WG) and WG members may share the WD within their own countries.
- Comments received from participating WG members are used to revise the WD which can again be shared within each WG member's country This procedure is followed until a consensus is reached by the WG members on the WD.
- The WD is then presented to the subcommittee (SC) to be accepted as a committee draft (CD). Subcommittees are responsible for developing the standards within a defined area.
- The CD is distributed to all SC members as a CD for ballot on four options as follows: Approve as is as a draft international standard (DIS); approve as a DIS with comments; disapprove the CD as a DIS; and abstain.
- If two-thirds of the returned ballots approve the CD as a DIS as is and/or with comments, it is elevated to a DIS.
- The DIS is forwarded to technical committee members after necessary revisions have been made and the members may approve or disapprove it as an ISO standard.
- If approved, all necessary editorial changes are done and a final ballot is taken on the revised DIS now refereed to as final or FDIS. Passage of this final ballot results in an ISO standard.

In Table 4.2, we present the EMS standards as of today based on the revisions of 2004. We also attach the most recent dates associated with each standard.

*Table 4.2: ISO 14000 Series of Standards*

| Standard number/Date | Status |
|---|---|
| ISO 14001, 2004 | International Standards |
| 150 14004, 2004 | International Standards |
| ISO 14010* | International Standards |
| ISO 14011* | International Standards |
| ISO 14012* | International Standards |
| ISO 14021, 1999 | International Standards |
| ISO 14020, 2000 | International Standards |
| ISO 14024, 1999 | International Standards |
| ISO 14041, 1998 | International Standards |
| ISO 14031, 1999 | International Standards |
| ISO 14042, 2000 | International Standards |
| ISO 14043, 2000 | International Standards |
| ISO 19011, 2002 | International Standards |

*ISO 19011 on environmental management systems auditing replaces ISO 14010, ISO 14011, and ISO 14012 on guidelines for quality and/or environmental management system auditing.

Table 4.3 contains other proposed EMS standards at different stages of development. It is presented below:

*Table 4.3: Working Documents on EMS Standards* (Table is adapted from http://www.tc207.org/pdf/ISO14000series1.pdf)

| Standard number/Status/Date | Description |
|---|---|
| ISO/TR 14025, 2000 | Environmental labels and declarations. Type III environmental declarations. |
| ISO/DIS 14025 | Environmental labels and declarations. Type III environmental declarations. Principles and procedures. |
| ISO/TR 14032, 1999 | Environmental management – Examples of environmental performance (EPE) |

*Table 4.3: (Continued)*

| Standard number/Status/Date | Description |
|---|---|
| ISO/DIS 14040 | Environmental management – Life cycle assessment – Principles and framework. |
| ISO/DIS 14044 | Environmental management – Life cycle assessment – Requirements and guidelines. |
| ISO/TR 14047, 2003 | Environmental management – Life cycle impact assessment – Examples of application of ISO 14042. |
| ISO/TS 14048, 2002 | Environmental management – Life cycle assessment – Data documentation format. |
| ISO/AWI 14048 | Environmental management – Life cycle assessment – Data documentation format (Revision of ISO/TS 14048:2002). |
| ISO/TR 14049, 2000 | Environmental management – Life cycle assessment – Example of application of ISO 14041 to goal and scope definition and inventory analysis. |
| ISO/NP 14050 | Environmental management – Vocabulary. |
| ISO/TR 14061, 1998 | Information to assist forestry organizations in the use of EMS standards (ISO 14001 and ISO 14004). |
| ISO/TR 14062, 2002 | Environmental management – Integrating environmental aspects into product design and development. |
| ISO/DIS 14064-1 | Greenhouse gases – Part 1: Specification with guidance at the organizational level for quantification and reporting of greenhouse gas emissions and removals. |
| ISO/DIS 14064-2 | Greenhouse gases – Part 2: Specification with guidance at the project level for quantification, monitoring and reporting of greenhouse gas emission reductions or removal enhancements. |
| ISO/DIS 14064-3 | Greenhouse gases – Part 3: Specification with guidance for validation and verification of greenhouse gas assertions. |
| ISO/WD 14065 | Greenhouse gases – Requirements for greenhouse gas validation and verification bodies for use in accreditation or other forms of recognition. |

Definitions: DIS – Draft International Standard; TR – Technical Report; TS – Technical Specification; AWI – Approved Work Item; WD – Working Draft; and NP – New work item Proposal.

ISO 14001 is considered the core standard because it is the only standard with specified requirements that firms must meet in order to achieve certification. A firm can therefore be audited on ISO 14001 standards. All the other standards listed in Table 4.2 are guidelines to help implement ISO 14001. These standards are not required for certification and a firm may not be audited on their basis. We shall briefly discuss the core areas covered by the ISO 14000 standards.

## Environmental Management Systems (EMS)

The core elements of an environmental management system (EMS) are presented in ISO 14001. The core consists of requirements that a firm can be audited on for certification. It deals only with environmental management standards and does not consider performance issues. To help implement EMS, ISO 14004 offers general guidelines on principles, systems and supporting techniques.

The involvement of top management is necessary when developing an environmental policy. The environmental policy must consider the environmental impacts of the activities, products or services of the firm. Management must be committed to continuous improvement efforts, and develop and implement plans for pollution prevention. It is important for management to ensure compliance to environmental legislation and regulations, and also to other regulations that the firm may already be committed to. This may involve establishing communication links with various interest groups. There must be an established framework to review environmental objectives and targets and the environmental goals of the firm must be documented and effectively communicated to all employees. The public should also be made aware of the environmental policy of the firm. Thus, top management has the responsibility of making the public aware of its environmental policy.

## Planning

A firm must develop a plan to help it achieve its environmental policy. Components of the plan are environmental aspects; legal and other requirements; environmental objectives and targets; and environmental management programs. Environmental aspects deal with procedures that the firm maintains to identify the environmental aspects of its activities, products or services. The firm makes an assessment of these impacts and determines its control over them and their expected impacts on the natural environment. The significant impacts must be considered in setting up environmental objectives. This information should be updated over time. It is a dynamic process that requires the firm to continuously monitor its environmental influence and impacts on the natural environment and update the available information as needed.

With regards to legal obligation and other requirements, it is the responsibility of the firm to be aware of the legal requirements it must comply with. It should maintain procedures to enable it to access such obligations that are applicable to the environmental aspects of its activities, products or services.

The firm must have environmental objectives and targets. These should be consistent with the environmental policy and commitment to pollution prevention. It is important that documentation is maintained at each relevant function and level within the organization. The objectives and targets should be cognizant of the legal and other requirements that the firm subscribes to, its significant environmental aspect, its technological options, financial, operational and business requirements as well as the views of other environmental interest groups. The targets should be measurable and specific and may be used to achieve the environmental objectives within a specified time-frame. Environmental management programs are the operational procedures to achieve environmental objectives and targets. They involve a breakdown of responsibilities for achieving objectives and targets; actions to be taken; resource allocation; and time-frame.

## Implementation and Operation

To effectively implement the environmental management program, the firm must develop the necessary capabilities and support mechanisms. This involves a well-structured organizational process where job responsibilities and authorities are well defined, documented and communicated. Resources needed to implement the program must be provided and management must be involved to ensure system viability and assess the performance of the program. A major aspect of implementation and operation of the environmental program is training awareness and competence. Competence may be developed through education and training.

It is important that trainees are aware of the requirements of the system and potential consequences of departure. Thus, appropriate training should be available. Communication is also an important aspect of implementation. The firm should have procedures for responding to relevant communications from external interest groups; and procedures for both internal and external communications. Like in many of the ISO standards, documentation is very important and could be either in paper or in an electronic form. However, there must be full document control procedures. Implementation must also deal with operational control of activities that are done under specified conditions. Suppliers and contractors should also be made aware of the procedures of the firm. The firm should have procedures to respond to emergency situations. This involves plan to respond to emergencies and procedures for accident prevention. These plans should be revised when an incident occurs and should be periodically tested.

## Checking and Corrective Actions

This step requires the firm to be able to measure, monitor and evaluate its environmental activities. This requires the firm to be able to monitor and measure key measures of performance, track operational performance, operational controls and objectives and targets. The monitoring process is only effective if the program complies with laws and regulations. Corrective and preventive actions may also be necessary when there is non-conformance. Auditing is conducted to assess conformance and proper implementation of procedures. A report is made available to management for review.

## Management Review

This requires the firm to review and continually improve its environmental management system in order to improve the overall environmental performance. Periodic review by management will ensure suitability, adequacy and effectiveness; address the need for policy changes or any other changes of the environmental management system; and documentation of the review.

We shall now focus on the other aspects of the ISO 14000 series.

## Environmental Auditing

This provides the standards that may be used for environmental auditing of the firm. ISO 19011 offers the general principles for environmental auditing. This standard deals mainly with the objectives and scope of the auditing, professionalism of the auditor, procedures, criteria, reliability and reporting.

It is the firm that commissions the auditing and states the scope and objectives of the auditing. Auditors however, are expected to be fair and avoid conflict of interest. They must also have the required professional skills and experience to enable them to fulfill these important responsibilities. Information obtained from auditing and the resulting report should be kept confidential unless the firm approves disclosure. Auditors must also follow documented and well-defined methodologies to carry out their auditing of a firm's environmental program. There must be consistency of auditing reports. In other words, other competent environmental auditors should be able to independently, reach the same conclusions. Auditors should also recognize that they are working with sample information and must therefore include some levels of uncertainty in their audits. All auditing findings should be communicated in writing to the firm.

## Environmental Labeling

The aim of environmental labeling is to reduce the environmental impact that may be associated with the consumption of goods and services. This serves several purposes:

- Labels are used to provide information on the environmental impact of a product or service and the consumer is made aware of that.
- The information content of the label may affect the purchasing behavior of the consumer.
- When purchasing behavior is influenced by the information content of the label, market shares may subsequently be affected.
- This will affect the attitude of manufacturers or firms who will respond to consumers' needs if they intend to remain competitive and increase their market share.
- There will be fewer burdens associated with the product or service.

## Environmental Performance Evaluation

This deals with a measure of performance. The ISO Sub Committee (5C4) defines it as a process to measure, analyze, assess, report, and communicate an organization's environmental performance. It is intended as a tool that assists company management in understanding environmental performance; determining necessary actions to achieve environmental policies, objectives, and targets; and communicating with interested parties.' (Block, 1997, p. 17). EPE focuses on three major areas: management systems, operational system and the environment. The management systems aspect deals with peoples' management. People within the organization take actions that may have an impact on the environment. There is a need for procedures and practice guidelines that relate to the management of the organization's environmental aspect. Operational system deals with process management. Here, the emphasis is on the transformation process to produce goods and services. Attention is given to the process itself in terms of equipment and physical structures and the materials and energy that are used to produce goods and services. The environmental emphasis is to focus organizational attention on its potential impact on all aspects of the natural environment. The organization is to assess the influence of its management and operational system performance on the environment.

## Life Cycle Assessment

The emphasis here is to evaluate manufacturing efficiency. Inventory analysis is employed to compile relevant inputs and outputs of a production system.

## Terms and Definitions

The aim is to co-ordinate the terms and definitions used by the various sub-committees and their work groups.

## Competing through Environmental Management Systems

Clearly, businesses are paying attention to ISO 14000 series of standards. Adherence to these standards can help an organization to be more competitive and increase its market share in a market environment that is increasingly focusing on 'green' products. The Standards Council of Canada (1997) in its publication provides a list of reasons why many companies are interested in adopting an internal environmental management system. These reasons are:

- Reduction of liability/risk;
- Improvements of a company's image in the area of environmental performance and compliance with regulatory requirements;
- Pollution prevention and energy/resource savings;
- Insurance companies' unwillingness to issue coverage for pollution incidents unless the firm requesting coverage has a proven environmental system in place;

- Better resale value of a company's property assets;
- Desire to profit in the market for 'green' products;
- Improved internal management methods; and
- Interest in attracting a high-quality work force.

These factors serve as a motivating force for companies to adopt the standards. We must also add that increased consumer awareness and the activity of environmental interest groups have greatly influenced attention on the environment. Consumers are now concerned about the environmental quality of the product and purchase decisions are influenced by environmental issues. It is a business and marketing strategy for organizations to achieve certification in environmental management system to show their responsiveness to environmental management.

## Revisions of ISO 14001

ISO 14001 which is the core of EMS was originally adopted in 1996 but was revised and adopted in 2004. The revision was intended to make ISO 14001 more user friendly by clarifying some of the statements in the 1996 document. It was also aimed to align ISO 14001 to the popular quality standards ISO 9001 and to establish clear association between the different segments of EMS, performance measurement, and the role of top management. This greater focus on alignment with ISO 9001 highlights the importance of quality imperative by emphasizing on Deming Plan-Do-Check-Act and continual improvement efforts. The revision also removed some of the vagueness in the original wordings of ISO 14001 by being specific on how some organizational environmental goals may be achieved. Munro and Harral [2006] classified these revisions into five "interpretative paradigm differences" as follows:

> - Communication – This deals with getting everyone on board to achieve the organizational environmental goals. They note that the greatest challenge is the increased detail that must be communicated to top management. The inclusion of "internal" communication in addition to external communication is emphasized.

- Documentation – There are changes in definitions, scope, and documentation requirements. Some of the definitions were borrowed from ISO 9001:2000. For example, organizations need to show that their auditors are competent. This concept of competence of auditors is derived from ISO 9001: 2000. Also, more succinct definitions of continual improvement and EMS audit are presented. Documentations have also been prepared to be easy to understand and also to demonstrate the significance of environmental aspects. There should also be documentation of results of periodic evaluations and monitoring of compliance.

- Competence – The definition of *competence* is still a gray area. Competence may vary from situations and challenges but organizations need to demonstrate by defining measures of competence. This could play major role not only in environmental auditing but may have labor and legal implications since competence extends to anyone that performs a task for the organization or rather, the entire value chain of the organization. The need for independent auditors is also emphasized.

- Performance focus and evidencing – The emphasis here is on measurement of objectives and targets and the need to see continual improvement as a "recurring" process and not a one-time thing. Resources must be readily available to support environmental goals.

- Legal and other requirements – There is need for a new level of awareness and this would require additional resources. There should be added emphasis on policy, objectives, or targets with resources devoted to them.

## Implementing ISO 14001

As we mentioned above, ISO 14001 is the core standard and it is the only standard that a firm can be audited on for certification. We also listed and briefly discussed the four core elements of ISO 14001 as environmental policy, implementation and operation, checking and corrective action and management review. In order to implement ISO 14001, an organization must go through these elements in a step-by-step procedure. These core elements are actually motivated by the Shewhart Cycle popularized by Dr W. Edwards Deming and now widely known as the PDCA (plan-do-check-act) cycle. The PDCA cycle is commonly used in implementing quality management programs. We shall use this cycle to show how these core elements of ISO 14001 can be implemented.

*Plan* — the planning stage requires the organization to develop an environmental policy. The environmental policy is akin to developing a mission statement that will detail the organization's roles, objectives, goals, and vision with regards to environmental performance. The objectives and targets specified in this statement must be realistic and achievable with the resources dedicated to attaining the environmental policy. Environmental policy is the motivating force of the organization's environmental management system. The organization however can only plan when it has relevant information. It needs to know its history, the nature of its business, and the mode of its interaction with the natural environment through its organizational activities. Thus, there is a need to have information and knowledge on 'environmental aspects.' The environmental impact of the organization's activities on the natural environment should be estimated, considered and used in setting environmental objectives and goals. The business or organization must also know the legal and regulatory requirements that guide its operations and how it is expected to comply with them. With this knowledge base and top management commitment, achievable objectives and targets can be developed and appropriate resources devoted to their attainment.

*Do* — this involves implementation and operation. Once the environmental policy is known, it is broken down into actions to be taken and responsibilities duly assigned to members of the organization. Necessary training is offered to sensitize and make members of the organization aware of the environmental policy, and to develop the needed competence on environmental management issues. They are also trained and made aware of the need to document their procedures. Emphasis is also placed on operational control and emergency preparedness and response.

*Check—Act* — the check stage involves monitoring the entire procedure and obtaining feedback. In the EMS document, it is referred to as checking and corrective action. The essence of this step is to evaluate outcomes of key performance measures and see if they meet expected standards or targets. The monitoring is done on a regular basis so that deviations from expected targets can be detected early. The targets or standards may be based on compliance required by existing legal and regulatory requirements. When the system is detected as not meeting these standards, corrective actions can be taken promptly. The compliance requirements are part of the environmental policy so there is a target to aim for. The act stage is included in this step because actions are taken as the situation may warrant solving impending problems such as system's deviation from expected norm.

The fourth core element of ISO 14001 is management review. This requires top management to be involved as an active participant of environmental management system. This is necessary, because certain actions or decisions can be taken at the top management level. Top management is required to review the EMS to ensure its continuing suitability and effectiveness. This review may lead to changes in environmental policy. For example, the original policy may not be adequate given some organizational transformation or process changes that may have taken place or it may not have been effective. Management will then require a revision of the environmental policy or development of new environmental policy that will align with corporate objectives and goals. The environmental policy drives the EMS and the organization's overall environmental performance so it is important that top management takes charge of this step. Once this step is completed, the process continues.

The implementation process offered here is generic and does not relate to any specific industry. It is a stepwise procedure that has to be taken irrespective of the industry.

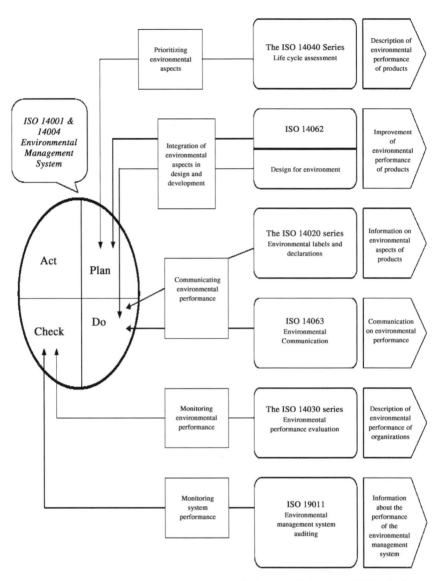

*Figure 4.1: ISO 14000 Model in an Integrated Plan-Do-Check-Act (PDCA) Cycle*

Figure 4.1 (adopted from http://www.iso.org) shows the ISO 14000 Model in an integrated Plan-Do-Check-Act (PDCA) cycle. The PDCA cycle was made popular by Edward Deming in quality management literature and has since been applied to several areas of study. Important information to derive from this is that the ISO 14000 model relies on effective planning, performance measurement, and monitoring. It is an ongoing process that seeks continuous improvement to achieve environmental quality. The effective implementation of these guidelines is essential to achieving maximum benefits and this would require effective planning. You would also notice that environmental management systems auditing is now represented as ISO 19011. This new standard replaces the previous ISO 14010, 14011, and 14012 and deals with guidelines for quality and/or environmental management systems auditing and not just guidelines for environmental auditing as in the previous standards. Thus guidelines for quality and environmental management systems auditing are unified.

## The Consumer and ISO 14000

The issue of standardization is of interest to consumers worldwide. Standardization ensures best practices and consistency in the delivery of products and services. It eases conformance to established guidelines and helps the regulation of products and processes. Consumers are protected from inefficient products and processes that are unsustainable. The quality of the environment and earth's limited resources are efficiently utilized.

Sustainability is not just of importance to consumers but to all stakeholders such as suppliers, manufacturers and vendors. We shall itemize some of the benefits of sustainable practices:

- Safer, healthier and environmentally friendly products are needed to improve the quality of life and productivity. Productivity as a measure of the economic wellbeing of a nation is enhanced when employees are safe and healthy. Environmentally sound products help to achieve the goal of increased productivity. One of the problems facing industrialized nations today is the increasing cost of healthcare and health insurance. Some of the health-related problems are induced by environmental pollution. Briggs estimates that 8-9% of total disease burden may be associated to environmental pollution and this figure is even higher for developing nations[8]. Yet, this percentage of total disease burden may be underestimated because of long latency times, difficulty in linking a pollutant to a single disease and multiple exposures to different pollutants. Major sources of environmental pollution include unsafe water, poor sanitation, poor hygiene and indoor air pollution. Why some of these may require basic hygiene practices however, industrial pollution contributes significantly in creating unsafe drinking water, poor sanitation, and poor air quality. For example, in many developing countries, there are few guidelines on factory locations and waste management. In such places, dumping of wastes and pollutants by manufacturers in streams and rivers and the lack of control on the emission of pollutants to the air pollute both the sources of drinking water and air. Standardization plays a role by specifying guidelines for best practices, sharing best practices worldwide, and educating regulators on standards to check for. The worldwide focus on best practices also compels manufacturers to carefully review and adopt ecologically friendly practices.

- In a global economy, it is important to have a level playing field. Consumers demand higher quality and quality extends to the role of the product on the environment. Consumers worldwide expect to get the same consistency of products and understand the need for safe and clean environment. They also participate in the green movement and would prefer manufacturers that are environmentally conscious. When a global company leaves its home base to compete in a new environment, it expects to meet exactly the same standards. By standardizing worldwide operations, the cost of operation and production is significantly reduced and high quality products that meet environmental needs can be delivered to customers at competitive prices. ISO standards facilitate international trade. By

[8] Briggs, D., "Environmental pollution and the global burden of disease," *Br Med Bull.*, 68: 1-24, 2003.

developing consistent standards, global companies can compete effectively by understanding the rules of the game. It would not matter if the company is based in Tokyo, Japan, or New Delhi, India, these companies do understand that there is a single world market that has to be catered for. Their products and services are evaluated using the same standards and their ability to compete effectively is dependent on their ability to satisfy these established standards and practices. So a sound business management practice would require knowledge of the guiding environmental management practices. The quest to meet and exceed these standards has made companies to become more innovative and find ways to turn environmental practices into profits. The cases of Kodak single use camera and Xerox remanufacturing practice show how corporations can be environmentally responsible and yet achieve high profitability.

• With the global economy, manufacturers are now dealing with global supply chain. Many manufacturers outsource part of their productions to other countries where cost of production is cheap, yet core competencies are available in such countries. Therefore, a manufacturer of aircrafts like Boeing may outsource the manufacture of wing flaps to Italy and expect to meet the same high quality and attain the same environmental standards. The attainment of these standards give consumers confidence that no laws are circumvented. In the past, multinational corporations relocated operations to countries where environmental laws were relaxed but today, they are joining in the effort to help such countries develop their environmental standards. Furthermore, with many of these countries as member nations of ISO, it becomes easier to develop consistent environmental standards worldwide. Using ISO 14000 standards and guidelines requires an evaluation of the value chain in order to support environmental protection and resource conservation efforts. This process helps in improving efficiencies and productivity. To effectively evaluate the value chain, the supply chain network becomes a critical component of this entire process. Many manufacturers have realigned their strategies with that of their supply chain to benefit from the global efficiencies these practices may lead to. So the issue is no longer being able to supply the cheapest cost but also being able to satisfy the standards and the reputation that the manufacturer wants. Thus the manufacturer and his team of suppliers work as team and share information on how to improve both product and environmental quality. Innovation is therefore critical in achieving both environmental performance and economic growth.

- Consumers in poorer countries stand to benefit from regulations since they could gain from the knowledge that exists in industrial nations. Poorer countries can benefit from this knowledge base without necessarily investing their resources on research and development to establish their own set of environmental laws.

- Green products create choices for consumers. Today's consumers are educated and have access to a wider range of information and database. They are able to make decisions that are rooted in their social and value systems. Consumers' perceptions of quality may be broader than the general definition of product quality and may focus on issues of social responsibility, integrity and trust [Madu and Kuei 1995]. Such focus on social and value systems are often associated to green issues. Consumers tend to perceive conformance to environmental standards as an aspect of organizational social responsibility function. Consumers today have a wide range of products and services to choose from and environmental issues are increasingly factored in making such decisions. Adhering to internationally accepted standards as outlined in ISO guidelines attest to an organizational conformity to established standards and elevates the organization above its competitors that may not demonstrate this mark of achievement. Companies that embark on environmental quality improvement efforts meet the needs of their stakeholders. They appropriately respond to the environmental challenges and develop a reputation of being stakeholder-focused. This will help create a business image and reputation that may transcend into increased market share and thereby higher profit margins.

- The use of ISO 14000 encourages environmentally sensible and conscious practices. This would also help to minimize ecological debts. According to Claude Martin, chairman World Wildlife Fund (WWF), "We are running up an ecological debt which we won't be able to pay off unless governments restore the balance between our consumption of natural resources and the Earth's ability to renew them,"[9] It is clear that a major problem is to be able to balance consumption of natural resources and the ability to renew the resources. While it is not always feasible to renew all resources, however, the use of ISO 14000 could help in responsible practices and in identifying sustainable practices that can extend the useful life of nonrenewable resources.

- In the past, different countries maintained different environmental standards. These standards were not universally accepted and were often contradictory. Such independent standards complicate international trade, regulation and monitoring, and do not protect global consumers. Today, the universal standards as achieved through ISO simplifies worldwide regulation, present the same view of environmental standards to all stakeholders, and assure consistency in achieving the standards. They facilitate international trade and ease entrance into new markets by foreign corporations. Consumers stand to benefit from competition, increased employment opportunities, and the quest by competing companies to be the best and produce world-class products and services.

**Environmental Action Box**
**A case study on Polaroid Corporation**

Polaroid Corporation is a good example of a company that adopts best practice environmental management program. Polaroid, which produces dozens of products, has adapted its product development process known as Product Delivery Process (PDP) to integrate aspects of design for environment (DFE). This integration ensures that environmental burdens are carefully evaluated and a life cycle assessment is conducted to ensure that the production process complies with the company's environmental goals. Polaroid enshrines

---

[9]   Fowler, J., "Group warns on consumption of resources,"
http://news.yahoo.com/news?
tmpl=story&cid=624&u=/ap/20041022/ap_on_sc/plundered..., October 22, 2004.

its environmental goals in its corporate mission thereby making this a major part of its operation. Even in the mist of restructuring, its CEO J Michael Pocock states in the 2002 sustainability report as follows

> *"At Polaroid, we understand that preserving the environment is not only a corporate social responsibility, but also an economic imperative. We safeguard the health and safety of our employees to show our commitment to responsible employment practices and to preserve the integrity of our business. But we also operate our plants to ensure that we protect the environment for our customers."*

This statement identifies some key challenges facing corporations today. They are concerned with the issues of social responsibility, integrity, competitiveness, and the role of their stakeholders. Companies today compete by listening to the "voice of the stakeholder." Employees and customers make up a segment of the stakeholder group and businesses cannot survive in an adverse environment. Polaroid however, has a long history of paying attention to the preservation of environment and views the environment not only from the social perspective but also as beneficial and key part of its business strategy. Its long history of environmental stewardship has led to some remarkable achievements such as the reduction in carbon dioxide emission, reduction in toxic releases, reduction in solid non-hazardous waste generation, and reduction in total energy usage. These reductions have made its products more environmentally friendly and competitive. Polaroid's sustainability program is not mere goals to achieve. They are also measurable. Polaroid uses a list of performance measures such as annual environmental compliance scorecard, third party audits and SARA Toxics Release Inventory. The key performance indicators involve looking at energy, water, toxic release inventory, waste generation, greenhouse gases, air emissions and compliance. The practice objectives of Polaroid focus on the following environmental management practices[10]:

---

[10] "2002 Sustainability Report," http://www.polaroid.com, downloaded on October 25, 2004.

| Environmental Practice | Yes |
|---|---|
| Reduced consumption of virgin materials through product or process redesign | X |
| Procurement of goods with recycled content | X |
| Recycling of solid waste | X |
| Recycling of hazardous waste and toxic materials | X |
| Water conservation | X |
| Energy conservation | X |
| Source reduction or risk reduction of toxic materials | X |

These practices are achievable and can be monitored.

One of the greatest challenges facing corporations today is the issue of outsourcing. Companies tend to focus on areas where they have core competence. By outsourcing, there is possibility that some of the guidelines maintained by a company may not be adhered to by vendors that supply to it. Outsourcing is an important part of many manufacturing and marketing operations. Taking a product stewardship requires that corporations must be responsible for their product and their supply chain network. Polaroid outsources the production of some of its major products such as digital cameras and instant cameras to companies across the globe notably Scotland and China. It is important that a systemic and global view of environmental management is taken in the production process. Polaroid has adopted stringent evaluation process of vendor pre-qualification to ensure that its environmental standards are maintained. Among the issues considered by Polaroid are those included in its Supplier criteria as presented below[11]:

---

[11]  Table is adopted from the 2002 Sustainability Report.

| Areas covered | Yes/No/Not Applicable | Comments |
|---|---|---|
| Necessary environmental permits | Yes | |
| Physical evaluation of facility | Yes | All suppliers are reviewed by a survey; site visits occur for "high risk" suppliers, such as chemical manufacturers. |
| Cooperative development of environmentally preferable materials, products and processes | Yes | On a case-by-case basis, Polaroid works with some suppliers to balance environmental responsibility with quality and cost issues. |
| Materials/energy efficiency | No | |
| Use of chemical in manufacturing | Yes | Among other areas, vendors are specifically asked about their use of ozone-depleting chemicals. |
| Chemical contained in product | Yes | |
| Product packaging | Yes | |
| Generation and disposal of waste | Yes | |
| Compliance | Yes | |
| Encouraging use of environmentally preferable materials in suppliers' processes and products | Yes | Processes are reviewed to ensure no ozone depleting chemical use. This review includes consideration of packaging that is recycled and/or recyclable. |
| Sharing company knowledge on environmentally preferable processes with suppliers | No | |
| Child Labor | Yes | Would disqualify vendor |
| Forced Labor | Yes | Would disqualify vendor |

Polaroid's lead in environmental management is remarkable. It was among the ten US companies that developed a set of guidelines between 1992 and 1993 to guide companies on how to publish corporate environmental reports. The group published its guidelines in 1994 as Public Environmental Reporting Initiative. The aim of the guideline is to serve as a tool for organizations to voluntarily produce a balanced reporting of their environmental policies, practices and performances. Polaroid endorsed the guidelines of the Coalition for Environmentally Responsible Economies (CERES) in 1994 and has been at the forefront of encouraging other companies to adhere to environmentally responsible practices worldwide. The principles of CERES are adopted in Polaroid's annual reports.

## A Case Study on the Dutch Airline KLM[12]

KLM is dubbed the first airline to achieve ISO 14001 certification. KLM adopted a pro-active environmental policy to respond to its changing business and political climate. The major thrust of its environmental policy was to achieve noise reduction and the safe disposal of hazardous waste. Major environmental issues of concern to airline operations include fuel and energy saving, noise reduction, emissions and wastewater, and waste separation. KLM's approach focused on planning and prioritization, employees' roles and responsibilities, providing resources and monitoring environmental management system (EMS) implementation. Influencing the attitudes of line management was instrumental in gaining their commitment and achieving environmental care. KLM achieved success by winning the support of its employees. Employees' involvement and commitment were instrumental in developing a system for continuous improvement through measurement and evaluation. Employees were supported with training on environment-related functions, and

---

[12] This case is adapted from Huiskamp, U., "KLM's ISO 14001-certified environmental programme takes off," *ISO Management Systems*, October 2001, 26-31.

environment orientation courses for strategic management. Employees were motivated to participate in the goal and communication of the environmental policies to stakeholders and listening to the stakeholders' voices were important in achieving sustainable policies. Today, KLM achieves environmental efficiency by

- Recycling newspapers
- Elimination of low-turnover goods in its duty-free selection on board to reduce weight and increase fuel efficiency
- Substitute fabric that does not require dry cleaning is now used in making flight-crew uniforms. This helps to reduce the amount of toxic dry cleaning chemicals emitted into the air.
- Optimal water requirement for a flight is maintained to reduce the weight on board and maximize fuel efficiency. KLM reports a savings of 1.6 million kilograms of fuel in 2000. Its fuel efficiency was 20% higher than the average European airline.

## Conclusion

In this chapter, we have discussed the ISO as an organization and its quality standards. We focused mainly on ISO 14000 series which are now emerging as the environmental management system standards for organizations. ISO 9000 and ISO 14000 series formats are similar in that they both require the development of frameworks to either manage product quality or the environment; they focus on continuous improvement and documentation procedures; and they rely on training and top management involvement. They, however, differ in content. ISO 9000 series focus much on customers and their needs but ISO 14000 series focus on 'environmental stakeholders' rather than customers. The organization must pay attention to the needs of this larger group that may not necessarily be its customers but have a stake in the natural environment. The need to attain sustainable development and protect our natural environment from degradation and pollution is a universal problem that is now transcending beyond business operations. So corporations of the future must pay attention to the need for sustainable development. We also looked into the revision of ISO 14001 in 2004 and noted that the major changes from the 1996 version focused on five areas: communications, documentation, competence, performance focus and evidencing, and legal and other requirements.

The increased number of environmental disasters such as the explosion of the Union Carbide's pesticide production plant in Bhopal, India, in 1984 and the Exxon Valdez oil spill in Prince William Sound, Alaska, has brought much focus on corporate responsibility. Madu [1996] points to the emergence of new environmental laws meant to regulate the operations of businesses and protect the natural habitat. With the avalanche of local, regional and national laws that corporations are expected to comply with, the cost of doing business in an environmental-sensitive society is increasingly high. It is important and cost effective to have international standards on the environment that can guide businesses and help them meet their corporate and social responsibility functions to the society at reasonable costs. ISO 14000 series of standards can achieve such a goal if they are widely accepted and adopted by member nations. Certification of corporations as meeting such standards can help reduce tensions and suspicions that often exist between communities and corporations. This will effectively enhance the image of corporations. Two case studies on Polaroid and KLM were discussed.

## References

Briggs, D., "Environmental pollution and the global burden of disease," *Br Med Bull.*, 68: 1-24, 2003.

Munro, R.A., and Harral, W.M., "The ISO 14001:2004 Revision," *Quality Digest*, February 13, 2006.

Madu, C.N., and Kuei, C-H., Strategic Total Quality Management, Quorum Books, New York, New York, 1995.

Madu, C.N., Managing Green Technologies for Global Competitiveness, Quorum Books, Westport, CT, 1996.

# Chapter 5

# Environmental Planning

A key issue in achieving sustainable development is the ability to manage human impact on the natural environment. Clearly, the scale of environmental pollution and degradation that are of major concern is that generated by industrial wastes through the creation of products and services. If such industrial wastes are not curtailed, sustainable development will not be achieved. Strategic planning as discussed in this chapter will examine the key facts both from managerial and technical perspectives on how sustainable development can be achieved through efficient environmental management. Specifically, we explore organizational strategic planning, competitiveness and the concepts of industrial ecology to understand how they influence sustainability.

While many of us are beginning to accept the need for sustainability, we must also not lose sight of the primary objective of a business or an industrial organization. Businesses have as their core objective to maximize shareholders' wealth. This objective may often be perceived by management to be in conflict with the goal of environmental protection. The bottom-line is what top management understands very well. Thus, to attract and sustain management's interest in environmental protection strategies, they must also be exposed to the potential benefits of such strategies. It is through thorough strategic planning of environmental issues that management would come to understand the value of environmental protection. To some companies, this is now a mute issue since they have already started to reap the benefits of sustainable development. However, more improvements can be achieved and it is necessary to continue to reinforce the value to organizations of sustainable development.

Environmental management is a corporate-level issue and demands the attention of top management. Many of the strategies required to achieve sustainable development may involve organizational restructuring, adaptation to new organizational culture, capital investment, and long-term planning. These aspects of planning are beyond the horizon of middle management and will need corporate-level involvement to be successful. Furthermore, being environmentally sensitive is a responsive strategy to both the needs of the customers and to the actions of competitors. Thus, it pays to be environmentally responsible. Also, the emergence of new laws and regulations in several countries is making it much harder for polluters to operate unnoticed. Corporations that have the vision to respond proactively to environmental demands will become competitive and have early entrance into new socially responsive markets. The cost of environmental pollution is high. Companies that pollute do not only suffer the high environmental penalty costs imposed by regulators but are also responsible for clean up costs and may in addition, suffer loss of market share from incensed public reacting to the company's environmental record. Typical examples include Exxon Corporation during the 1990 oil spill at Prince William Sound, Alaska. Customers as a result of this accident repudiated its credit cards. Union Carbide also suffered public humiliation and discontent when its chemical plant exploded in Bhopal, India in 1984.

## Strategic Environmental Management

Environmental management is a core company value and not simply a public relations ploy [Grant and Campbell 1994]. Corporations realize the business potentials that can accrue from responsible environmental management. Consumers are gradually shifting their priorities and supporting products and services that are environmentally friendly. Grant and Campbell [1994] note that the environment can be integrated into the corporate culture in several ways such as:

- Expanded innovation and productivity,
- Better environmental performance.
- Improved bottom line.

- Better handling of volatile environmental controversies.
- Enhanced credibility and trust.
- More employee involvement in community relations.

These approaches on how to incorporate environmental management in to the corporate culture also offer guidance on what is needed in the environmental front to achieve competitiveness in the growing environmental market. Clearly, the end result for profit making organizations will be to improve the bottom line. However, to achieve such goal, the corporation needs to develop more efficient and innovative system or processes to enable it achieve the demanding societal environmental goals. By meeting the demands on the environment, the organization builds credibility and trust and is able to deal with volatile environmental controversies. In return, market shares may be gained, cost may be reduced and the bottom line may be improved. Employees also develop a sense of pride and joy in their organization and become integral members of their communities. Responsible environmental management is a win-win strategy that benefits the business enterprise, consumers, employees, and the society as a whole.

When the company is environmentally sensitive, it enjoys the support and goodwill of its community. The company becomes more efficient and cost conscious. The cost of poor environmental quality is controlled thus enabling it to invest in innovation and Research and Development. This may lead to the generation of new and improved products and services that will have less demand on material and energy consumption. The company is also able to position itself as a leader in its industry as it continues to gain market shares.

Strategic planning requires adopting a vision for the future. The world community has a stake in the survival of the natural environment and with new laws and regulations passed by the different nations and the world communities, businesses must heed to the need to protect the natural environment. Manufacturers are recognizing that in order to continue to be competitive, they must be environmentally conscious. Thus, the traditional manufacturing strategy and philosophy must change to respond to our changing environmental needs. There is greater desire now than ever to produce environmentally friendly products. These products must make efficient use of limited natural resources, create less waste, and have less demand on material and energy. The focus is to increase or maintain value why reducing input requirements for a product. The reduction in input results in elimination of wastes, less dependence on nonrenewable resources, less dependence on energy and material, and more efficient use of technologies. To achieve this, the entire production system must conform to the new environmental standards. The goal here shifts away from the end-of-pipe management approach where the aim is to treat waste at the end of the entire process but rather, to prevent the waste from being incurred in the first place. A cradle-to-grave approach is therefore taken starting from the idea conception stage through product design, production, and usage and disposal stages. Environmental management and protection is an ongoing process that spans through the life of the product. It is supportive of business ventures and it should not be seen as a costly venture but as a means of improving the long-term profitability and survivability of the firm. Businesses must see themselves as partners in this because if the source of their input is polluted or depleted, then the future of the business will also be affected. Imagine the impact that shortage of pulp would cause the paper industry. However, through effective planning, the availability of the resources could be protected through effective usage and recycling efforts, replanting of the forestry, and conservation.

## Environmental Planning Framework

Planning for environmental protection will take the same traditional approach as any strategic planning framework. Madu [1996] presented an environmental planning framework referred to as the strategic cycle or system transformation process. This figure is adapted and presented below.

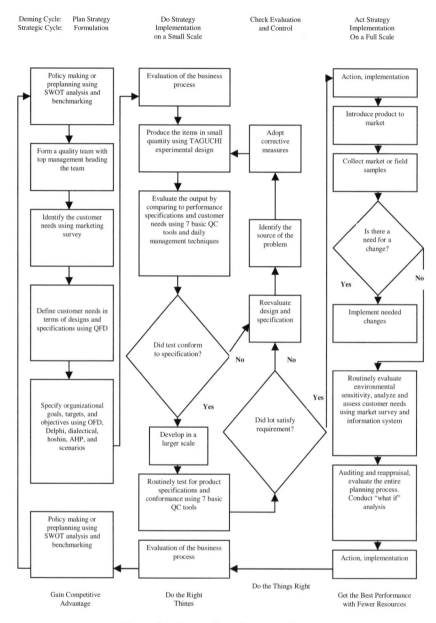

*Figure 5.1: System Transformation Process*

We shall discuss each of the key steps or phases in this planning process below. It is shown in this figure, that environmental planning process is a never-ending cycle that is akin to the popular Deming cycle in quality management. Dr. Edward W. Deming popularized the use of the Plan-Do-Check-Act in quality management and it is shown here that the strategic cycle for environmental management follows basically the same steps. Therefore, environmental planning can be broken down along these four major phases as we describe them below:

## Plan (Strategy formulation)

Plan can be referred as the strategic formulation phase. It is a critical element of any organization's strategic formulation. The organization's vision on how to deal with its customers and its extended and external environment is made at this point. Let us take for example, the case of new product development. The company decides through its research and development that a new product to satisfy a particular need has to be developed. This new product is not without its consequences to the natural environment. Developing the product will require the use of natural resources, materials and energy and its usage may create other environmental burden such as emissions to the atmosphere or disposal to landfills. At this stage, it is important to adequately plan on how to deal with these environmental burdens. There are several things that an organization can do to minimize the environmental burden of the product. One of the approaches to be followed may include forming an environmental quality group or a team that will comprise of all major stakeholders. These stakeholders will conduct a critical assessment of the product from a design, production, and disposal point of views and contrast each of the needed elements with substitutes to identify which options may have the less environmental burden. The formation of a stakeholder team brings in people with diverse worldviews and experiences and may lead to proliferation of new ideas on how to best satisfy customer needs. This approach also allows for a more thorough SWOT (strengths, weaknesses, opportunities, threats) analysis where the manufacturer does not view SWOT only internally but also externally. In other words, environmental burden and concerns of the general public about the product could be factored into the designing of the product. When the manufacturer is able to integrate the views, concerns and values of the general public in product design, it becomes more capable of seeking alternative production systems and product substitutes that may have less environmental impacts. It also becomes more capable of working with its supply chain to make sure that the entire supply chain network is adequately involved in achieving its overall goal of environmental quality. With manufacturers increasingly outsourcing different stages of their product development to different suppliers and vendors, it is important that the entire supply chain network work towards a common goal. Irrespective of who the vendor or supplier may be, the manufacturer must take a cradle-to-grave approach of its product.

The concept of SWOT analysis is deeper than just the manufacturer identifying its internal strengths and weaknesses or opportunities and threats from a competitive point of view. A major strength that a manufacturer can enjoy is its stakeholder's goodwill. The general perception by the stakeholder that the manufacturer is environmentally conscious by itself is a competitive weapon that could translate into increased market share. Threats could also be seen from the point that competitors have more environmentally friendly products and opportunities might rise from the fact that existing products do not meet the environmental needs of consumers. SWOT analysis could in fact, help an environmentally conscious manufacturer to identify environmental market niche that are not being satisfied by competitors.

Environmental management planning would benefit from the use of existing management tools. One of the issues involved in using stakeholder teams in planning product design and management is that an avalanche of issues may be raised regarding the interaction of the product with its natural environment. Also, several substitutes for the different components required for the product may be suggested and would need to be evaluated in terms of effectiveness, quality, cost, and environmental burden. A company could conduct a comparative assessment by using existing management tools to determine how best to satisfy customer needs through design. One of the strategies that have been effectively used in the quality management literature is to apply the quality function deployment popularly known as QFD. QFD is effective in matching customer needs to design needs. QFD drives the company by pushing the design team to identify "hidden" customer requirements and offering ways to satisfy such requirements. Its use also helps to identify those factors that are important to customers so that the design phase does not miss out on the critical feature that a product should have. Furthermore, it could help the company to conduct a competitive and technical evaluation of itself relative to its competitors and to how well it is achieving specified target values [Madu 2000, 2006]. The point is that QFD helps facilitate the planning and production of products and services that meet customer expectations and will therefore help ensure that manufacturers adequately integrate customer concerns about environmental burden in product design and production.

Developing a product to satisfy the environmental needs of the consumers is not all that easy. It requires a clear and objective assessment of the impact all stages of product development will have on the natural environment, thereby demanding that a thorough life cycle assessment should be conducted. Life cycle assessment is often difficult to conduct since it is difficult at times to prioritize different environmental impacts. For example, in comparing cotton and softwood pulp, it is difficult to determine which of these two creates less environmental burden. Further application of management tools would require the ranking of the environmental burdens that may be identified by the stakeholder team in terms of relative importance. This may also consider situational and structural differences that may affect alternative choices. The use of the analytic hierarchy process to achieve this is discussed in Chapter 6.

Once the planning phase is done, the next phase is to execute the plan. The execution phase is the "Do" stage.

## Do (Strategy implementation on a small scale)

This phase involves the evaluation of the plan. It may involve pilot studies, developing and testing small-scale models, and computer simulations. The objective here is to mimic the real life product through modeling so to be able to address "what if" scenarios that may arise. This phase involves testing different substitutes or alternative designs and collecting samples for further analysis as regards to their environmental burden or impacts. This stage provides critical information that could be used to narrow down the potential number of substitute components for a product or the number of alternative designs that are being compared. In the end, only those components and alternative designs that meet the specified threshold are kept and further evaluated on other factors such as resource availability, cost, quality, and compatibility with other products or services provided by the manufacturer.

## Check (Strategy evaluation and control)

Test marketing on a sample group could be conducted to obtain further information on how to enhance the product and its environmental quality. This is often difficult to implement since some of the environmental impacts may not be readily available and may take few years to realize. However, a full-blown implementation could be risky if proper product testing is not conducted. The testing phase is meant to see if the established environmental goals and standards are being satisfied. If not, the sources of the problem should be identified and rectified before a full implementation should be considered. Also, even when the established standards are being satisfied, it is important to match the prototype product to competitors' products. This will help to determine market acceptability and product strengths that could be used in devising marketing strategies for competitive advantage. This testing procedure is therefore, an effective means of acquiring new information on the product through comparative assessment of its environmental features against the products of competitors.

## Act (Strategy implementation on a full scale)

The product after undergoing all the necessary corrective measures should then be introduced into the market on a large scale. The stakeholder team would be supportive of the fact that adequate measures have been taken to ensure that environmental burden created by the product has been limited through effective planning. However, even with all good intent, there is no certainty that the product will not create any unwanted environmental burden. A monitoring program needs to be implemented to track the product through its life to detect manageable environmental burdens. Also, with improvements in new technology and the availability of new information, it may be possible to continuously improve on the product or replace it or some of its components with more environmentally friendly substitutes. The concept of product stewardship requires the manufacturer to take responsibility for its products through their life cycle and to continuously seek methods to improve the environmental quality of the products. When significant environmental changes are needed, continuous improvement may no longer be effective and a complete overhaul and perhaps redesign of the product may be required. In this case, reengineering of the entire process from design through production and usage may be required. The key however is that the planning process is a never ending process that constantly updates itself using new information and feedback obtained from the natural environment. It is an open system that seeks to find the best way to design and produce environmentally friendly products and services. Continuous improvement while helpful with small changes that may affect the product may become inefficient when there are significant environmental changes. When such occurs, reengineering of the entire product and production process may be required.

## Understanding Environmental Problems

The environmental strategic planning framework presented here would be incomplete if it has no means of identifying an environmental problem or burden that may exist with a particular product. To facilitate the identification of these problems, we propose the use of a popular tool in quality management known as the fishbone diagram or the Ishikawa diagram or the *4m*. Basically, this fishbone looking diagram is based on the idea that every problem can be deciphered into four parts namely man, machine, methods and material. We believe this to be also true in managing environmental problems. We use the fishbone diagram to analyze environmental problems associated with paper production.

- Man—the role of the labor force in creating waste and environmental pollution is rather obvious. Human error often results in the misuse of limited natural resources such as raw materials and energy in producing needed goods and services. Also, some of the major environmental accidents have been attributed to errors in human judgment, poor supervision and training, and often lack of sensitivity to environmental needs. Errors in carefully evaluating and deciding on alternative substitutes or production processes to limit environmental burden can also be attributed to management problems which are human in nature.

- Material—environmental burden can be created when less efficient material is used. For example, in the paper industry, the use of virgin pulp against the use of post consumption paper would lead to unnecessary exploitation of forestry thereby diminishing the limited forestry resources. Likewise, the decision to use composted wastes in farms against synthetic fertilizers will not only enrich the soil but prevent the erosion of top soil and the excess buildup of nitrogen in the soil. Recycling and reuse of materials have also helped greatly in limiting environmental burden attributed to material usage. As we have illustrated in the case of Kodak single-use camera, material conservation can be achieved by remanufacturing and reuse. The use of fossil fuels can also be limited by finding renewable energy resources. Efficient manufacturing and production strategies also reduce both material and energy inputs.

- Machine—environmental burden may be created when equipment or machinery malfunction or fail to produce within specified tolerance limits. As a result, more scraps or rejects are incurred thus placing more demand on the need for both material and energy resources. Furthermore, emission to land, air and sea could be affected if the machine is improperly serviced or fails to meet emission standards.

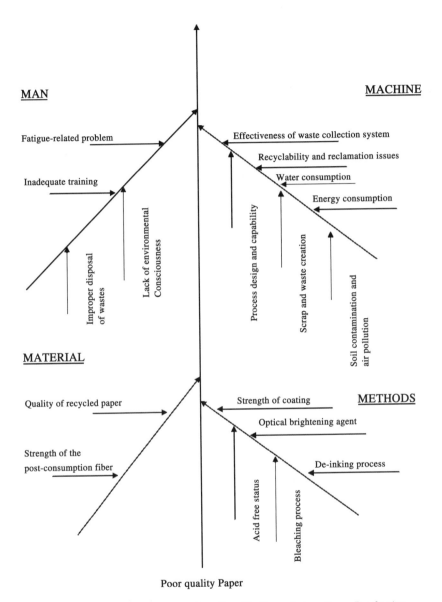

*Figure 5.2: A Fishbone Diagram for a Specific Example i.e., Paper Production*

- Methods—the design strategy detects the technique or method of production and could significantly influence the output of wastes and energy consumption. Different design strategies such as design for environment, design for manufacturability, and design for recyclability are widely used to limit environmental burden. A diagram is presented below to illustrate the use of the *4m* in analyzing environmental impacts.

From this discussion, it is apparent that an efficient planning process must include consideration of man, materials, machine, and methods and how they could potentially influence environmental burden. The environmental impact of each product or service should be evaluated on the basis of these *4m*s to better project its environmental burdens. This diagram should also be used to evaluate alternative or substitute products in parallel. It gives a clear view of all products and improves understanding of both the strengths and weaknesses of each design strategies. It could be used as a basis not only for selecting the product with less environmental impact but also for improving production systems by benchmarking a process that appears to be more efficient and environmentally friendly.

## Organizational Culture and Environmental Planning

Planning cannot be effective without a change in organizational culture. Everyone has to do things differently. This will require a total system overhaul with top management taking active part and showing support for environmental management initiatives. A total new culture has to be built and this culture will require attitude and value changes from employees. Top management must commit not only time but resources in terms of factory modernization by adopting new and more environmentally friendly processes, education and training of employees to teach them the importance of environmental-quality standards and their role in meeting the environmental goals of the company, and better relationship with suppliers and vendors to ensure that they follow corporate environmental guidelines and policies. The training should emphasize the importance of self-regulation, governmental standards and legislation, and the influence of environmental interest groups and how to work with the different parties to achieve sustainable manufacturing practices. Employees should be acquainted on both the short- and long-term environmental forces and the role of their companies to mediate and abate some environmental causes.

When senior management is aware of the importance of environmental quality planning, it begins to analyze the buying patterns of consumers as they relate to the environment. By understanding that the buying patterns of consumers are significantly shifting from an emphasis on direct product quality to also include environmental quality, senior management realizes the necessity to develop a new organizational mission and vision for the company based on environmental goals. This is followed by the development of specific strategies and programs to meet customers' needs and to achieve the goal of company wide total environmental quality. The increase in the demand for organic food products is a case in point. In 2003, the US and EU accounted for 95 percent of worldwide retail sales of organic food products in the tune of $25 billion[13]. While the European organic markets are more mature, the US is steadily catching up. In the 1990s, the average growth rate of organic retail sales in the U.S. was 20 percent per year and this pace is expected to continue to 2005 and then slow down to the range of 9-16 percent per year through 2010. This trend is also expected to overflow to other industrial sectors.

Senior management needs to encourage process and organizational transformation and redesign. These actions involve not only process transformation but also cultural transformation.

As shown in Figure 5.3, customers' needs are holistic and there must be a balance between environmental protection and product satisfaction. Ideally, the consumer will prefer a high-value product that is also environmentally friendly. These emerging needs of the consumer present new challenges, opportunities and threats to the company. The company

[13] Dimitri, C., and Oberholtzer, L., "EU and U.S. Organic Markets Face Strong Demand Under Different Policies," *Amber Waves*, February 2006.

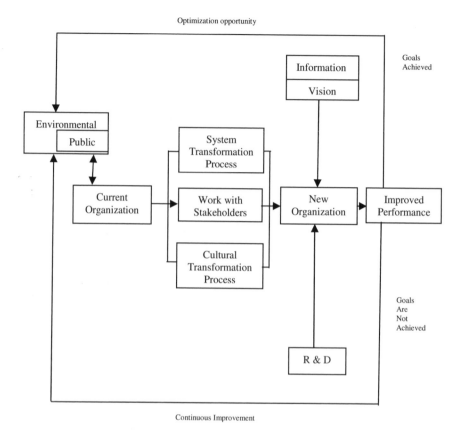

*Figure 5.3: Transformation to A Green Organization*

would need to undergo a process of change in order to adapt to the new environment. However, these challenges pose new opportunities that could be exploited to gain new market share. If it is however ignored, it becomes a significant threat to the survival of the firm.

There are three main areas that are impacted by the process of change. The first is the system transformation process, which deals with the process of transforming inputs into outputs. Here, adequate control standards must be in place to ensure compliance with stringent environmental laws. The problem is not just producing products or services that meet environmental standards but also ensuring that the transformation process itself meets the same stringent conditions. Thus, the product or service, as well as its transformation process during every stage of production, should meet the same environmental constraints to be in full compliance.

Companies today have a large network of suppliers, distributors and vendors they work with. The concept of product stewardship requires that the manufacturer take a cradle-to-grave approach of its products through its life cycle. This requires establishing modalities to effectively work with partners or affiliates within the supply network to ensure that consistency is maintained in environmental standards and that the product is effectively monitored throughout its life.

Criteria should be established to measure the level of effectiveness of the supplier in meeting environmental standards. Inevitably, the manufacturer's environmental quality and cost of products will be affected adversely if the materials provided by the suppliers do not conform to environmental standards.

## Listening to the Voice of the Stakeholder

In developing our planning framework, we emphasized the importance of the stakeholder team. This team would be made up of active participants with multidisciplinary background and different worldviews. The team provides valuable information to the organization on its wants and needs. A customer-focused organization can only be effective when it listens to its customers and designs and develops products and services that meet their needs. We have mentioned the importance of using the quality function deployment (QFD) tool in designing products and services that meet customer needs. It is imperative that such needs are identified by talking and working with the customer. We have purposely referred to members of such customer group as stakeholders since the group is made up of not only those who purchase the products or services but all those that have a stake at the outcome of such products and services. For example, the emission of carbon and other greenhouse gases to the atmosphere affects all. It is not an issue that is left alone to the consumers of the product. Likewise, the filling of landfills with solid wastes that are not biodegradable affects the quality of our natural environment and therefore, concerns everyone. It is therefore more pertinent that a stakeholder rather than a customer group be used in assessing the impact of corporate activities on the environment.

It is noteworthy that many corporations are already joining the trend of involving environmental interest groups as stakeholders in their strategic plans. This partnership is helping in developing more sustainable services that will enjoy wider support from the general public. Some of the examples include the collaboration between the fast food chain McDonald's and the Environmental Defense Fund (EDF) in switching from polystyrene containers to paper wrap for its food products. Similarly, the utility company Pacific Gas and Electric Company often collaborates with the Natural Resources Defense Council. The Conservation Law Foundation also worked with New England Electric System to develop a twenty-year strategic plan that will focus on improving the use of renewable resources for power generation. This trend is continuing and shows increasingly that manufacturers are responding to the needs and wants of the general public by developing strategies to achieve sustainability.

Increasingly, visionary corporate leaders are adjusting their mission statements to include an ethical value system as part of their organization's social responsibility. To effectively achieve such corporate missions, organizational changes may be required and made part of the corporate environmental strategies. We shall look at some of these below:

- Top management commitment—Top management commitment is the most important factor in ensuring that corporate environmental strategies are successful. Top management shows the lead, which other employees must follow. Top management involvement highlights the importance of sustainability and good environmental practice. Furthermore, it is top management that has to allocate and commit resources to ensure that the goal of minimizing environmental burden is effectively carried out. Sound environmental practices often would require the re-engineering of the entire organization. For example, decisions would have to be made to modernize factories and replace mundane processes, training of workers would be initiated, suppliers may have to be retrained or replaced to meet environmental goals, vendors and distributors may need to be coached along, and management style may need to be changed to create more openness, sharing of ideas, and involvement of important stakeholders in environmental decision making. All these may lead to changes in the organization's culture and such changes need to be managed. Decisions involving such drastic changes rest with top management. Thus decisions on environmental sensitivity are strategic and can only be effectively considered at the top management layer. Any other level of management would lack the required authority to make decisions of this magnitude let alone implement them. It is also the responsibility of top management to ensure that good environmental practices are companywide efforts that do not stop with the production of goods or services but also become part of the organization's life.

- Organizational vision and mission—The organization needs a vision that would serve as the guiding philosophy to help it to accurately plan and predict its future. This of course is based on understanding its strengths, weaknesses, opportunities and threats (SWOT). SWOT analyses help the organization to better handle its future and understand its competitive environment. The vision articulates where the organization is and where it intends to be. Good visions require good imagination from top management and a deep understanding of its working environment. Good vision will separate better-managed organizations from their competitors as they identify new trends and market niches for the future. Vision helps the organization to plan ahead of its rivalry and position itself in its environment and prepare itself to respond to changes that might arise. It helps to restructure and redesign an organization so that it can cope with changes that may affect it. Finally, vision is based on good articulation of customer needs and wants, challenges in the marketplace, and understanding of competitors reactions and responses to the dynamic market environment. Being environmentally sensitive and responsive is a challenge that organizations face and must respond to in order to remain competitive.

- Change management—Clearly, introduction of cleaner, leaner and more environmentally friendly culture in an organization would require re-engineering of the organizational structure and of course, organizational cultural changes. The value system of employees as well as their perceptions of nature would have to change to align with the new vision and missions of the organization. Adherence to the new philosophy of environmental quality improvement should be obvious in their new role as the guardians of the environment and should be prevalent in all their activities within the organization and in the organization's extended environment. Employees therefore, need to be trained and sensitized to be able to identify potential sources of waste and pollution and to be able to control pollution or waste from their sources. Management of change is not easily accomplished. It would require changing life-long habits and adopting new culture that is supportive of sustainable development. Furthermore, it requires that management empowers employees so they can have more leverage in making decisions and taking actions regarding their work and their role in the workplace. Top management is the key in making this transition from traditional practice to the new environmental practice culture smooth. It needs to lead by example and listen to the voices of both the customer and the employee and grant them participation in workplace decision making.

- Designing for the new environment—Effective management of the environment depends a great deal on designing. Products and services have to be designed to achieve high levels of efficiency. We have noticed increasingly the use of recycled materials or products in paper, electronic, automobile, computer industries and others. Recycling strategy helps to prolong the life of materials and substances that are used, reduce the energy demand required in excavating new materials, and yet assuring the efficiency and the effectiveness of the new product. Achieving sustainable design that meets the needs of the customer requires that design engineers work with the customer to understand and design products and services that meet its needs. This approach challenges the traditional practice of engineers designing products as they see fit for use and then shoving them down to the customer to be consumed. The customer today is an educated consumer who is often aware of his or her needs and shows sensitivity to the natural environment that ultimately is involved in the systemic production process. Designing products or services to meet the challenges of today and tomorrow require that a thorough environmental impact assessment of all the components of the product or service be conducted by tracking the entire production process and the product life cycle to ensure that the final product creates the minimum environmental burden. For example, decisions on which alternative material or component to use in a product should be based on estimating each material's lifetime impacts on the natural environment and not simply on one or few direct impacts that are quite obvious. A design team that is made up of important stakeholders with multidisciplinary backgrounds should therefore be used to effectively analyze all pertinent information.

- Competitive benchmarking—A competitive benchmarking approach needs to be adopted. In other words, the organization has to transform itself as a learning organization and be able to learn and adjust its strategies from world-class organizations. In other words, companies can learn from other organizations that have best-in-class sustainable manufacturing programs irrespective of their industry. For example, Kodak is a leader in product recycling as evident from its single-usage camera, Xerox is a leader in remanufacturing, and LL Bean is a leader in packaging which helps to trim waste significantly. Learning from these world-class organizations can help transform existing practices and make them more efficient. Knowledge of the corporate practices of these leading companies could help to establish achievable environmental goals and targets.

- Environmental cost—A more comprehensive cost assessment methodology should be developed. One of the problems facing environmental management is that environmental costs are often undermined or underestimated. Top management should be made aware of environmental costs especially since manufacturers now must take a cradle-to-grave approach of their products. Product stewardship could create an enormous economic impact on the company since liability extends through the entire life of the product. All the facets of cost such as internal, external, appraisal, and prevention costs must be evaluated. When top management is away of the array of costs involved and their impact on the competitiveness of the organization, it becomes more willing to support efforts toward sustainability. One cannot overlook the importance of profitability to corporate enterprises and the need to often quantify practices to top management in terms of cost-benefit analysis. It is therefore of utmost importance that any environmental assessment program should have a major component on environmental cost analysis.

- Corporate image and social responsibility—Organizations have a stake in ensuring environmental quality. They operate to serve the general public whose perceptions of the performance of the company in their extended environment can influence their purchasing decisions and reactions to the organization. Customers have publicly demonstrated against utility and gas companies they perceive as polluting the environment and have even gone as far as boycotting their products. Corporate image can significantly be hampered if the company is perceived in a bad light and this perception could limit its ability to provide its intended social responsibility function to the society. Apart from their role in providing jobs, companies also have the responsibility to protect and improve the quality of the natural environment. Companies that fail to deliver these important services to their communities are often entangled in litigations and poor public relations with their communities.

- Strategic information management system—sharing of information is necessary in the effective management of the environment. Corporations today deal with a supply chain network of vendors, distributors, suppliers, and customers. Every group in this network participates to effectively deliver high quality goods and services to the customer. They all need to meet the established goals and standards of the organization and this can only be possible by sharing information with them and providing support to each group so the same target is achieved. The different members of the supply chain network can also obtain independent information that can be shared with any member of the network. The ultimate goal should be to improve the current level of performance using accurate and timely information. Strategic alliances should also be formed with major stakeholders with the openness and flexibility to communicate information to all groups with the intention of achieving the organizational goals.

## Conclusion

In this chapter, we looked at the planning issues involved in environmental management. We note the importance of top management involvement and commitment in ensuring effective environmental planning. Issues evolving around environmental planning may often involve organizational restructuring or reengineering and such drastic changes can only be made by top management involvement and commitment of resources to that effect. Therefore, environmental planning is a strategic responsibility that rests with top management. We have also discussed the need of using a Plan-Do-Check-Act cycle as a means to effect a new organizational change to achieve sustainability. This systematic approach to planning will diffuse the influence of sporadic changes without adequate planning. It is also important to note that in planning for environmental management, the root causes of environmental pollution, material waste and consumption should be identified as well as their potential effects. We noted that this could be done by classifying all causes into what we refer to as "*4ms*" notably man, machine, methods, and material. Understanding these *4ms* will help to isolate all the problems in any particular system. Finally, the need to listen to the customer and understand organizational cultural changes is underscored. We have identified several areas that top management must focus on to achieve its corporate environmental missions.

## References

Grant, A.J., and G.G. Campbell, 1994 "The Meaning of Environmental Values for Managers," *Total Quality Environmental Management*, 3(4): 507-512.

Madu, C.N., Managing Green Technologies for Global Competitiveness, Westport, CT.: Quorum Books, 1996.

Madu, C.N., House of Quality (QFD) in a Minute, Fairfield, CT.: Chi Publishers, 2000.

Madu, C.N., House of Quality (QFD) in a Minute, Fairfield, CT.: Chi Publishers, 2nd edition, 2006.

# Chapter 6

# Life Cycle Assessment

## Introduction

LeVan [1995] traced the history of life cycle assessment to 1969 and noted that the first life cycle analysis was conducted on beverage containers. The aim of this analysis was to determine the type of container that had the least impact on natural resources and the environment. This led to the documentation of the energy and material flows although the environmental impact was not determined. Since this initial work, LCA has been broadened to focus on inventorying of energy supply and demand for fossil and renewable alternative fuels. Thus, the focus of LCA is no longer inward with a concentration on the direct influence of the product but also outward to consider the energy and natural resources input. Also, the increasing concern about limited landfill spaces and the health risks associated with pollution have generated the need for a more holistic view of environmental impact assessment.

## Definition

There are two major definitions of life cycle assessment. These definitions are provided by the Society of Environmental Toxicology and Chemistry (SETAC) and the International Organization for Standards (ISO). Both groups have been active in developing guidelines for LCA. SETAC defines life cycle assessment as:

"An objective process to evaluate the environmental burdens associated with a product, process or activity by identifying and quantifying energy and materials used and wastes released to the environment, to assess the impact of those energy and materials uses and releases on the environment, and to evaluate and implement opportunities to affect environmental improvements. The assessment includes the entire life-cycle of the product, process or activity, encompassing extracting and processing raw materials; manufacturing, transportation, and distribution; use/reuse/maintenance; recycling; and final disposal."

ISO's definition appears in the ISO 14040.2 Draft: Life Cycle Assessment - Principles and Guidelines and is defined as:

"A systematic set of procedures for compiling and examining the inputs and outputs of materials and energy and the associated environmental impacts directly attributable to the functioning of a product or service throughout its life cycle." This goal is accomplished by the following steps:

- Compiling an inventory of relevant inputs and outputs of a system;
- Evaluating the potential environmental impacts associated with those inputs and outputs;
- Interpreting the results of the inventory and impact phases in relation to the objectives of the study.

There are three major components of life cycle assessment. These are: inventory analysis, impact assessment and improvement assessment. LCA is a way of making the manufacturer to take responsibility for its products. It induces a design discipline that aims at achieving more value for less where the definition of value is expanded to include the potential impacts of the product or service on the environment. The designer focuses on design option that is environmentally sensitive by evaluating the product's demand for limited resources, energy, and disposal requirements at every stage of the product's life. Emphasis is on potential environmental burdens, energy consumption, and environmental releases. The manufacturer also takes a product stewardship approach in evaluating the product, process, or activity. Environmental impacts include the expedition and use of limited natural resources, the pollution of the atmosphere, land, water or air, ecological quality (i.e., noise), ecological health, and human health and safety issues at each stage of the product's life cycle. We shall now, discuss the three stages of LCA.

## Three Components of Life Cycle Assessment

As we mentioned above, there are three major parts to life cycle assessment. These three parts are discussed below:

## Life Cycle Inventory Analysis

The aim of life cycle inventory analysis is to quantify energy and raw material requirements, atmospheric emissions to land, water and air (environmental burdens), generation of solid wastes, and other environmental releases that may result throughout the life cycle of the product, process or activity within the system boundary. These environmental problems affect the quality of the environment and in many ways; the public pays the cost of environmental burdens. The costs of these environmental burdens are often difficult to estimate since they are not all direct costs. Some of the costs may not even be detected until several years after the damage has been done. There are both economic and social costs that are involved. In order to conduct life cycle inventory analysis, we must associate the environmental burdens with functional units. In other words, it should be measurable. For example, we need to use a standardized measurement to quantify waste or raw material and energy consumption. We can for example, measure the carbon emission to the atmosphere in metric tons or the per unit weight of solid waste from a particular geographical location. The functional unit should provide information on the composition of waste both in terms of material type and relative weight [Kirkpatrick, 1999].

Inventory analysis is the thrust of LCA. The normal production process involves actually three main steps: inputs, transformation, and outputs. Each of these steps is a major source of environmental burden and environmental releases. By looking at each of these steps, the process of data collection for inventory analysis can be enhanced:

Input - The input stage involves the acquisition of raw materials and energy resources. Inputs can also come in the form of transfers from other processes. For example, a recycled product can be a source of raw material for producing new product or semi-finished product from a different production source.

Transformation - The transformation process normally deals with the process to convert the input into a desired output. The transformation process also involves energy consumption as well as information flow. Further, wastes could be created through the process as a result of systemic problems with the process itself.

Output - Output may be in the form of finished product, which is shipped out to the consumer, or semi-finished product that becomes input in another process. Also, at the output stage, two types of outcomes can be expected: products that meet the quality guidelines and those that fail the quality requirements. There is therefore, the potential that waste may be generated at this stage both in terms of raw material consumption and energy that is used to generate such wastes.

It is therefore important that these three stages of the production process be evaluated in order to generate an inventory of raw material usage and energy consumption as well as environmental releases.

In the quest for environmentally conscious manufacturing, one of the popular strategies today is to seek for better environmental alternatives. For example, polyethylene and glass, which one is more environmentally friendly? Or, should cloth diapers replace disposable diapers? Important information that is generated in life cycle inventory analysis is known as the table of impacts. This is a table that presents the impacts from the possible production of two materials. For example, we can look at the emissions and solid wastes generated in the production of 1kg of polyethylene and compare it to that for the production of 1kg of glass. However, such evaluation cannot really suggest to us which alternative is better without taking a systemic view of the entire production process. For example, in a study by Johnson [1994], he noted that cloth diapers will require more chemical releases and water usage for cotton while softwood pulp for disposable diapers will require more energy requirements. These two options: cloth diapers and disposable diapers create environmental burdens and it is difficult to compare the environmental burdens. So, how does one make a trade-off between these two fibers? There are therefore, a number of problems that make it difficult to conduct life cycle assessment. Some of these problems as they relate to inventory analysis were identified by Product Ecology Consultants [1999] and are discussed below:

## Problems with Life Cycle Inventory Analysis

- Boundary conditions - It is difficult to define the system's boundary. For example, how far should one go in identifying inputs and outputs that relate to a particular product or process? Based on SETAC guidelines, components that comprise less than 5 percent of the inputs should be excluded. This is however, problematic since it is based on the assumption that the 5 percent component in a product will not have a significant environmental burden. If we go by the ABC rule, there is the potential that a very small fraction of the components may indeed, contribute to the majority of the environmental burden observed. LeVan [1995] presented a good example by noting that the electricity used for particular activity may be a small part of the input. However, if such electricity is generated from a high sulfur coal plant, its environmental burden could be enormous.

- System boundary condition - There is also the possibility that the links to certain products may be traced infinitum. Kirkpatrick [1999] notes for example, that the production of polyethylene involves the extraction of crude oil which is transported in a tanker. The tanker is made of steel, and the raw material required for steel is extracted. If we continue, we can see a long product chain that grows larger and larger and becomes more complex to analyze. Thus, a line must be drawn on what constitutes the system's boundary. This will generally not include capital goods.

- Multi-product processes - Some processes are designed to generate multiple products. In such cases, it is not easy to allocate and assign environmental burdens and releases to the different products.

- Avoided impacts - When materials are incinerated, energy is normally generated. Such energy is considered an impact but also, saves impacts, as it will no longer be necessary to produce the energy or the material. These avoided impacts are similar to the impacts that would have occurred in the production of material or energy. They are also, deducted from the impacts caused by other processes.

- Geographical variations - This recognizes the fact that environmental needs may be geographically dependent. We shall present two examples one from LeVan and the other from Product Ecology Consultants. LeVan [1995] notes that in the time of draught in the U.S. Southwest, single-use disposable diapers would be preferred to home-laundered diapers. While these two create environmental burdens however, the need at the time is a prevailing reason for the choice of single-use disposable diaper. Product Ecology Consultants [1999] on the other hand note that an electrolysis plant in Sweden will create less environmental burden than in Holland because hydroelectric power is in abundance in Sweden.

- Data quality - Environmental impact data are often incomplete or inaccurate. The data can also become obsolete and the use of such data may lead to distortion. There is also a problem that some of the environmental burdens may not be known and there may in fact, exist no data on the environmental impacts.

- Choice of technology - Clearly waste, energy consumption, and material releases to the atmosphere can be linked to the type of technology as well as the maintenance of the technology. Poorly maintained vehicles emit more carbon to the atmosphere and so are poorly maintained manufacturing processes. Also, the precision of such processes may be questionable leading to more creation of wastes. Further, modern technologies may meet the new environmental laws and are able to control emissions and some environmental wastes and pollution.

## Life Cycle Impact Assessment

Life cycle impact assessment is a way of interpreting and aggregating the inventory data so they could be useful for managerial decision making. We mentioned above that one of the important information generated through life cycle assessment is the table of impacts. The table of impacts is however, difficult to interpret without further evaluation of the environmental impacts. It is important to evaluate the impacts by assessing their relative contributions to different environmental concerns. Kirkpatrick [1999] lists some of the impact categories that are often considered in environmental impact assessments. These are:

- Resource depletion
- Greenhouse effect (direct and indirect)
- Ozone layer depletion
- Acidification
- Nutrification/eutrophication
- Photochemical oxidant formation.
- Other areas that are less well defined were also identified as
- Landfill volume
- Landscape demolition
- Human toxicity
- Ecotoxicity
- Noise
- Odor
- Occupational health
- Biotic resources
- Congestion

All these areas pose environmental hazards and need to be considered in assessing environmental impacts. There are however problems with assessing environmental impacts. For example, the fact that data is often non-existent and even when data may be available, it may be difficult to accurately determine the extent of the damage to the environment. Another reason is that there is no standardized method of estimating or measuring environmental damages.

The aim of life cycle assessment is pollution prevention rather than pollution control. By identifying the major sources of pollution and their effects on the environment, efforts could be made through product design or product development to prevent waste and reduce the risks to the environment.

## Measuring Environmental Impacts

It is difficult to measure environmental impacts. One of the beginning steps is to first identify a complete chain of cause and effect. But as we noted in the system boundary condition, this could become very cumbersome making it difficult to effectively analyze the problem.

We recommend here, the use of Fishbone diagram to graphically identify the potential causes and effects. However, we must point out that there will be series of such diagrams since some of the causes or effects will need to be further analyzed. The use of Fishbone diagram offers great opportunity in breaking down the potential causes in to 4 parts or what is known as *4ms*: man, machine, material, and method as we discussed in Chapter 5. A sample of the fishbone diagram is presented as Figure 6.1.

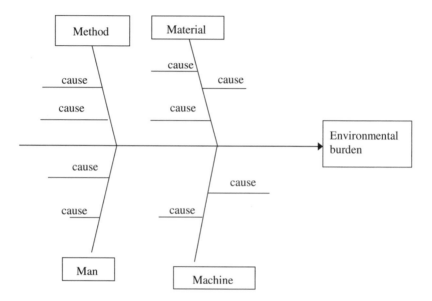

*Figure 6.1: Fishbone Diagram for Environmental Impact Assessment*

A similar method to the Fishbone diagram is used in Sweden to check cause and effect [Product Ecology Consultants 1999]. This is known as the EPS (Environmental Priority Strategy). The Swiss government on the other hand, uses the Ecopoint method, which measures the distance between the current impact and the target level. This is used to determine the level of seriousness of the impact to the environment.

SETAC presents the procedure for environmental impact assessment that involves three steps:

- Classification and characterization
- Normalization
- Evaluation.

However, only the classification and characterization stage has actually been implemented in practice. We shall discuss them with insight on how they may be improved.

## Classification and Characterization

The fishbone diagram provides a good method of identifying causes and effects. This stage calls for classification of all impacts based on their effects on the environment. Impacts could be grouped based on their contributions to the different environmental burdens such as Resource depletion, Greenhouse effect (direct and indirect), Ozone layer depletion, Acidification, Nutrification/eutrophication, and photochemical oxidant formation. Also, an impact can be classified in more than one area. Thus, an effect table may be generated that will be of the form shown below:

*Table 6.1: Effect Table for Impact Assessment*

| Emission | Quantity (kg) | Greenhouse | Ozone layer | Acidification | Nutrification / eutrophication | Photochemical oxidant formation |
|---|---|---|---|---|---|---|
| Nitrogen dioxide | | | | | | |
| Sulphur dioxide | | | | | | |
| Carbon dioxide | | | | | | |
| Carbon monoxide | | | | | | |
| Nitrogen monoxide | | | | | | |
| Effect score | | | | | | |

The contribution of each emission to each environmental burden is assessed and its effect score computed. As the intensity of effects may vary among the different chemical compounds, it is important to use a weighting factor. We suggest the use of a systematic weighting scheme such as the Analytic Hierarchy Process (AHP). A pairwise comparison of these chemicals could be conducted to determine their relative contributions to each of the environmental burdens. For example, carbon dioxide and sulfur dioxide, which has the most effect on greenhouse effect? Such weighted factors or priority indices obtained through the AHP could be used to obtain the effect scores. Thus, each quantity will be multiplied by the priority indexes generated through AHP and added up to obtain an environmental burden effect score. Clearly, there are many effect scores that must be generated. Emission for example is one area of environmental burden. Others are resource depletion, landfill usage, etc. The effects could all be compared as well as the relative importance of the different environmental burdens in a particular geographical area. This may form the basis for an informed decision on an environmental policy.

SETAC Workgroups have developed frameworks for life cycle impact assessment (LCIA). They identified the elements for conducting LCIA (Life cycle impact assessment) as classification, characterization, normalization, and valuation. Since SETAC is championing the effort in developing the methodology for LCIA, we shall briefly discuss the work of its two workgroups from North America and Europe. The group proposed four major elements for life cycle impact assessment framework. These are classification, characterization, normalization, and valuation.

The classification phase involves creating different categories for inventory results. This will help distinguish and group impacts for planning purposes.

The characterization phase involves the conversion of inventory results in a category into a category indicator. Equivalent categories are aggregated into a category indicator. The category specific models are constructed using a cause-effect diagram.

The normalization phase involves normalizing the category indicators by dividing them by a reference value. This helps broaden the scope of the interpretation of the data by comparing the different category indicators.

The valuation phase is based on developing a formal ranking of category indicator results across impact categories. The weights or rank order are subjectively determined.

The use of AHP can also provide an alternative approach to LCIA. We shall therefore, discuss the AHP concept below.

## Analysis of the Use of AHP

The Analytic Hierarchy Process (AHP) has three main components: goal, criteria, and alternatives. The goal is what is to be accomplished which in this example, is to select the most environmentally friendly product (i.e., glass or polyethylene). However, this decision depends on several factors denoted as criteria. These factors include the contribution of these products in creating environmental burden such as greenhouse effect, ozone layer depletion, energy consumption, and others. With each of these environmental burdens, there are several other sub-criteria to consider (i.e., emission of gasses that affect the ozone layer or cause the greenhouse effect). All these affect the decision on which of the two products to select. Figure 6.2 shows the hierarchical network structure of this decision making process.

The use of AHP as a decision tool has been widely published [Saaty 87, Madu & Georgantzas, 1991, Madu 1994, and Madu & Kuei, 1995]. The AHP is defined by Saaty as "a multi-criteria decision method that uses hierarchic or network structures to represent a decision problem and then develops priorities for the alternatives based on the decision makers' judgments throughout the system" (Saaty, 1987, p.157). The features of AHP that makes it applicable for application in life cycle impact assessment are the following:

- It allows for a systematic consideration of environmental problems by identifying all the major environmental impacts such as the greenhouse effect, depletion of the ozone layer, eutrophication, human or ecological toxicity as well as the factors that may influence them such as the emission of certain types of gases to the atmosphere, use of technology, consumption patterns, etc.

- There are many players when it comes to environmental issues. Environmental policies are not purely technical. There are many interest groups whose views must be aligned to develop sustainable environmental policies. The AHP makes it easier to consider all these different stakeholders in developing environmental policies.
- Its technique is novel as it deals with issues of consistency in decision making. Also, priorities generated through AHP can offer a good guide in reaching decisions on the relative importance of the different environmental options.
- Non-technical information can be combined with the more quantitative and scientific information on the environment to reach a decision.
- It helps to breakdown complex problems into levels of complexity that are manageable. As Figure 6.2 shows, the impact assessment problem can be broken down into the following parts: goal, criteria, sub-criteria and decision alternatives. This makes it easier to systematically analyze the problem.
- AHP helps to measure the consistency of the decision-maker. Although consistency does not guarantee quality decisions, however, all quality decisions are consistent.

Madu [1999] applied the AHP in analyzing the allocation of carbon emission to inter-dependent industries. The use of AHP as a decision support can help to clarify problems when comparing alternative choices since it will attach relative importance to different environmental impacts.

## Life Cycle Improvement Analysis

Life cycle improvement analysis is akin to the use of continuous improvement strategies. The goal in environmentally conscious

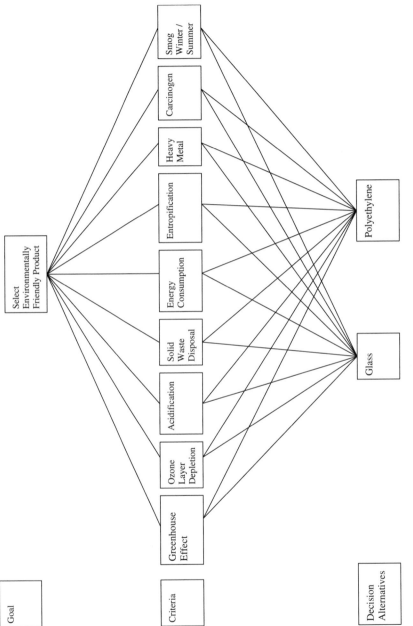

*Figure 6.2: An Example for Selection of Environmentally Friendly Product with AHP*

manufacturing should be to achieve "zero pollution." This can only be achieved if the entire system processes are continuously improved on. For example, the process of managing the life cycle of a product is a long and an arduous task that starts from raw material acquisition through manufacturing and processing to distribution and transportation, then, recycling, reusing, maintenance, and waste management. Each stage of the product life cycle involves creation of waste, energy consumption, and material usage. A life cycle improvement strategy is needed to identify areas where improvements can be achieved such as a product design that has less demand on material requirement i.e., the reduction in the size of automobiles, replacement of 8 cylinder engines with 4 or 6 cylinder engines; development of hybrid vehicles; enhanced consumer usage of products by reusing products or creating alternative uses for products, cutting down consumption of fossil fuels, participation in recycling programs, and so on. Life cycle improvement analysis will require the tracking and monitoring of the product through its life cycle to detect areas for continuous improvement. Life cycle improvement analysis has led to a lot of changes in design thus, the design for environment.

## Design for Environment

Design for environment is a design strategy that ensures the design of environmentally friendly products. This is done by paying attention to the importance of recycling, waste minimization, and reduction in energy consumption. We shall briefly discuss the popular design strategies in designing for environment.

- Designs for recyclability - products are now being designed for ease of disassembly so that materials and components can be recovered at the end of the product's life cycle and reused. This process extends the useful life of components in many products and limits the demand for virgin items. This may therefore, lead to the conservation of energy that may be needed to excavate new materials.
- Design for maintainability/durability - increased reliability of products ensures that they will have extended operational life thus

reducing the demand for newer generation of products. However, if it is difficult to maintain or repair products, there may be more demand on landfills for disposition of such items. Solid wastes will be created as well as energy wastes.

- Designs for pollution prevention - environmentally friendly products are now widely used in the production process to eliminate the amount of toxic and hazardous wastes that are being generated. Further, since products are designed for ease of disassembly and recycling, the amount of wastes generated is significantly reduced.

## The Use of Life Cycle Assessment

Life cycle assessment studies should be carefully used. There is the tendency for manufacturers to use such results for marketing purposes. However, any such claims as to which product is better than the other based on life cycle assessment studies may be misleading. There are inherent problems in doing life cycle assessment studies that make it difficult to use such studies for comparative purposes. Some of the reasons are identified below:

- Life cycle assessment studies cannot definitively state that one product is better than the other if we follow through the product's chain and study all the potential interactions of the product with the environment. It is difficult to make a decision on which form of pollution is better or preferred to another. For example, is the emission of toxic gases preferred to the use of fossil fuels? No form of pollution should be encouraged. Thus, our target should be to achieve "zero pollution." Until that goal is achieved, it will be incomprehensible to suggest that one form of pollution is better than the other.

- Life cycle assessment will be incomplete without the consideration of both quantitative and qualitative data. Further, there are cases where either data does not exist or may be incomplete. When we start integrating perceptions and other non-quantitative issues in the decision making process, it becomes difficult to come up with a non-subjective decision.

- Life cycle assessment is limited by both boundary condition and system boundary conditions. Thus, results generated may be geographically dependent rather than universal. So, a claim that one product is better than the other may depend on where the product is being used and may therefore, provide misleading information to consumers.
- Recycling is also a major source of complexity in life cycle assessment studies.
- The systemic structure of life cycle assessment makes it difficult since environmental processes vary in spatial (i.e., geographical) and in temporal (i.e., in time frame and extent) [SETAC Workgroup].
- There are also assumptions in life cycle assessment studies that there are no thresholds and that there exists a linear response between the system loading and the environment [Fava et al. 1991].
- Other potential problems include the fact that a solution to one form of environmental pollution may actually lead to another form of pollution. Typical example is the attempt to reduce the dependence on landfills for solid wastes. While this may reduce the pollution of air and groundwater supply, however, the burning of such wastes may lead to emission of energy to the atmosphere.
- Like the example we presented earlier, can one conclude that the use of cloth diapers is more environmentally friendly than the disposable paper diaper? Each of these presents an environmental burden.
- Data collection problems can plague the effectiveness of life cycle assessment. Sometimes, it may be difficult to correctly assess and collect information on all the inputs and outputs from a process. Further, life cycle analysis is greatly affected by the quality of the life cycle inventory analysis. The life cycle inventory analysis is the foundation for life cycle analysis and must be done right.

These drawbacks therefore, make it difficult to use LCA in certain ways. Many have suggested that the adoption of eco-labeling schemes may standardize the comparison of products in terms of their environmentally friendliness. Eco-labeling standards are currently popular in Europe and the Nordic countries as well as Japan but the US does not have a national program for eco-labeling. Eco-labeling is however, not without problems. A major concern is that third-party certification bodies may adopt different and confusing appraisal methods.

LCA however, can be used as a strategic planning tool in the following ways:

- It helps to identify which areas to focus on in order to achieve the environmental protection strategy. It is a proactive and a systemic way of looking at the company's products and services.
- The cradle-to-grave approach helps to ensure that manufacturers innovate on how to minimize wastes and eliminate environmental pollution. Further, by adopting a cradle-to-grave approach, emphasis is placed on the evaluation of multiple operations and activities throughout a product's life cycle to explore and manage potential sources of environmental pollution.
- The use of a systemic framework enables a functional analysis of the potential impacts of pollution on the environment. Thus, the different effects of energy consumption, material resources utilization, and emissions on the different environmental media such as air, water, and land, and waste disposals can be estimated in order to reach an effective decision on environmental protection.
- Emphasis is on optimization rather than sub-optimization. This is done by looking at a product in its totality by checking the consequences of its interaction and interface with the global environment. A product is therefore seen as interacting with the outer system. Feedback and information are received on how to improve the product. Thus, a product's quality is not simply measured by its ability to deliver its intended function but also by its ability to deliver such function at a minimum social cost to the society.
- Life cycle impact assessment can be extended as a means of benchmarking competitors by developing more environmentally friendly substitutes and improving the design for environment.
- Eliminating wastes, cutting down on pollution, and increasing the recyclability of components used in product design can reduce production costs.
- Customer satisfaction and loyalty can be improved by designing and producing products that meet customers' environmental needs.

## Strategic Planning for Life Cycle Assessment

Life cycle assessment can be seen as a strategic tool. It is a tool that will enable the manufacturer to understand the nature of his business and the needs of his customers. Critical questions that are asked in strategic planning include: What business are we in? Who are our customers? And who are our competitors? A well-designed life cycle assessment may enable a manufacturer to address these questions. First, what business are we in? Every business exists for a purpose and one common purpose for manufacturers is to provide goods and services. These goods and services must provide value to customers otherwise; there will be no demand for them. But, what is value? Customers' needs are ever changing. A few years back, there was not much concern about the degradation of the natural environment. Once a product meets high "quality" standards, it is expected to do well in the marketplace. Today's needs are different. Customers are concerned about environmental degradation resulting from depletion of limited natural resources, emissions to the different environmental media such as air, water, and land, and the influence of entire environmental burden on their quality of life. The pressure is on manufacturers to take a more environmentally responsible approach of their products and services. Hence the need for life cycle assessment. As manufacturers grapple with life cycle assessment, they must also come to understand that environmental issues are systemic in nature and are far reaching. Focusing on direct customers alone is too narrow. As a result, manufacturers should be concerned with stakeholders rather than customers. Stakeholders are active participants who are affected by the environmental burden and whose actions can also affect the role of the manufacturer in contributing to the environmental burden.

Manufacturers must strive to be the best or world-class performers in all their operations. This gives them competitive edge. Being the best is a selling point. As we noted earlier, sometimes, LCA is misused by manufacturers who tend to advertise claims that their products may be better than that of competitors based on life cycle assessment. However, there are obvious cases where a manufacturer may be making serious efforts to innovate and reduce its products' contributions to the environmental burden and the competitor sees no need to embark on environmental protection programs. Further, some manufacturers or service providers have achieved remarkable results like in the packaging industry or in digital rather than paper invoicing that they have become companies to be benchmarked. A holistic view of the environmental component of a product or service is not only economical in the long run as it reduces liability costs and other associated costs, but it is also strategic as it positions the manufacturer to compete effectively and expand its market base. While we have discussed the technical aspects of life cycle assessment such as life cycle inventory analysis, life cycle impact assessment, and life cycle improvement analysis, it is also important to look at the non-technical aspects that are strategic in nature.

It is crucial that life cycle assessment is an integral part of any organizational decision making process. The framework presented below has a focus on environmental management which life cycle assessment is a major part of.

## Strategic Framework for Life Cycle Assessment

This framework is broken into three major parts preplanning, evaluation or impact assessment and action implementation. Action implementation may also involve improvement analysis.

## Preplanning

The starting point for the framework is the formation of stakeholder team. Although most components of life cycle assessment may be scientific and objective in nature however, there are environmental perceptions and concerns that may not be effectively addressed. People are often concerned about potential impacts. Local conditions such as political and economic issues may influence perceptions. The use of stakeholders in decision making helps to ensure that the concerns of the stakeholders are considered and perhaps, integrated into the decision-making process. This will make it easier for the stakeholders to accept the final outcome of this process thus making it easier to adopt and implement the final decision. It also helps the manufacturer to expand its scope by considering environmental impacts that may have been neglected internally. For example, stakeholders may take issue with the recycling program or location of landfills.

Participation of the stakeholder team will help the manufacture to obtain information and feedback, and hear directly from the "voice of the stakeholder." Thus, the needs and concerns of the stakeholders are better defined and aligned with organizational goals. By working with the stakeholder team, the different environmental media of concern to the stakeholders can be identified and the relationship of the product manufacturing strategies to these media can be better understood through a life cycle inventory analysis of each product strategy. The analytic hierarchy process (AHP) discussed above can be used to evaluate the different product strategies or scenarios based on the life cycle inventory analysis and the concerns of the stakeholders. This will help to develop a portfolio of product strategies through which an informed decision could be made on potential strategies for adoption. However, there is the issue of cost and feasibility of some strategies. For example, some product alternatives may not be technologically or economically feasible. All these have to be taken into consideration in narrowing down the choices of effective product strategies. The AHP allows establishing priorities for the different product strategies. A systematic consideration of product strategies for implementation can be based on their priority assignments. Once a product strategy is selected, it is matched with design requirements. This phase is accomplished by using the quality function deployment (QFD) as a tool. The QFD if effectively used will enable the manufacturer to effectively match its capabilities to stakeholders need and also benchmark its design capabilities to that of competitors [Madu 2006].

## Evaluation or Impact Assessment

This stage involves the evaluation of the product design to ensure that all the significant needs of the stakeholders have been considered in the product design. A prototype or simulated product may be developed and its potential impacts on the environment and product performance estimated. The simulated impacts will be compared to established standards and targets from the preplanning stage. If the estimated impact from this stage does not conform to expectations, the product design and specifications need to be re-evaluated to identify the source of the problem. Corrective actions are then taken. If however, the simulated impacts conform to expectations, the product could be developed and a random sample of the product may be used for environmental test marketing. The problem with this stage is that environmental impacts are often not measured in the short-term and may take a long time to show up. However, the use of expert analysis could be helpful in evaluating the potential impacts of the products although there is no guarantee that these potential impacts may be fully estimated.

## Action Implementation/Improvement Analysis

The product is now introduced into the market after it has been certified as meeting the established standards. While the product is in the market, the process of data collection continues. Routine tests are conducted to ensure that the product continues to meet the environmental requirements. Further, the availability of new scientific information may suggest new and more environmentally friendly components that could lead to redesigning of the product to conform to such changes. Also, new legislatures may impose limits on emissions or other environmental burden that may demand a change in product design. Once the product is out in the marketplace, information gathering must continue to create an ongoing process of improving the environmental quality of the product.

The Life Cycle Assessment Framework presented here is a continuous loop where feedback is frequently being fed into the system. The importance of this feedback loop is to ensure that the framework is timely and able to respond quickly to environmental changes. It is a dynamic framework and regards the availability of new information as a necessity to improving the quality of life cycle assessment.

The framework is shown in Figure 6.3 below:

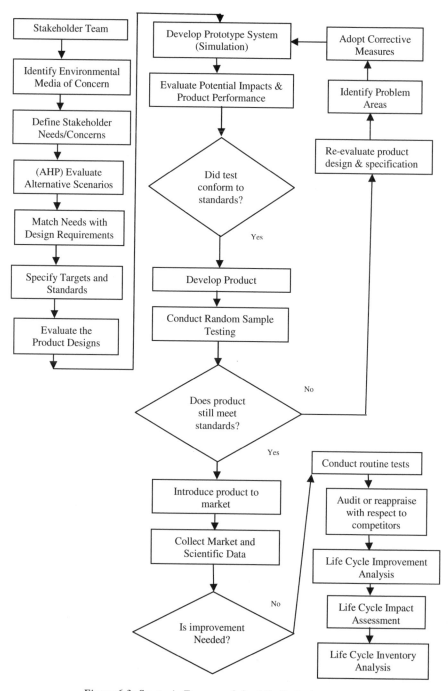

*Figure 6.3: Strategic Framework for Life Cycle Assessment*

## Life Cycle Cost Assessment

Life cycle assessment models must address the cost issues. If life cycle assessment is not economical and cannot translate to improvement of the bottom-line, it will be difficult to assure the compliance of manufacturers. One must therefore learn to speak the same language as manufacturers. That is, cost and profitability issues must be frequently addressed in conducting life cycle assessment studies.

Life cycle assessment cost is therefore, a method of evaluating costs that are associated with achieving environmental compliance. Such costs may include the conventional costs that may be incurred in implementing warranty programs (i.e., recalls of products that fail to meet environmental standards), social cost to the environment (i.e., costs of clean up and costs of decreased productivity from health and safety factors), liability costs (penalty and legal costs), environmental costs (i.e., environmental pollution costs and health-related costs), costs associated to loss of customer goodwill and negative campaigns against the manufacturer. These costs can be grouped into the four types of cost of quality accounting system introduced by Dr. Joseph Juran namely: internal, external, preventive and appraisal costs [Madu 1998].

- Prevention costs - involve the cost incurred by ensuring that environmental pollution is prevented. Some examples of this cost include the costs incurred by conducting life cycle assessment as shown in Figure 6.3, training, designing for environment, product review and preplanning, vendor selection and so on.
- External failure costs - these costs are incurred after the product has been manufactured and shipped out to the consumer. Environmental problems found are rectified and the product is returned back to the consumer. Such costs include costs of maintaining warranty, liability, recall, social, loss of customer good will, penalty/fines, complaints, personal and property damages and so on.
- Internal failure costs - These costs are incurred internally before the product is shipped out to the end user. Such costs include the costs of rectifying high levels of emission in the production

process, material usage and wastes such as scraps, energy consumption and expenditure, cost of inspection tests and retesting, and so on.

- Appraisal costs - This cost is incurred by ensuring that the process and the product meet the target environmental standards or to ensure compliance. Such costs include inspection, equipment calibration, product audit and design qualifications, conformance analysis, and monitoring programs to track wastes and pollution, and environmental quality audits.

When top management understands the cost of poor environmental quality, it will take seriously the effort to improve environmental quality and will therefore, pay attention to life cycle assessment.

## Environmental Action Box

## A CASE STUDY ON LIFE CYCLE ASSESSMENT

The case study presented here is adapted from LeVan [1995]. In her article, she discussed the work done by Franklin Associates Ltd. [1992], for the American Paper Institute and Diaper Manufacturers Group as well as the work by Johnson [1994]. These results provide insight for developing the case. However, we use the AHP to analyze the results. The data used here are partly hypothetical and partly estimated from the figures presented by Franklin Associates Ltd. in its studies. The focus of this case study is to conduct a comparative assessment of energy consumption, water requirements, and environmental emissions associated with the three prominent types of children's diaper systems: single-use diapers containing absorbent gels, commercially laundered cloth diapers, and home-laundered cloth diapers. The comparison is based on a usage of 9.7 cloth diapers per day and 5.4 single-use diapers per day. Environmental emissions here are an aggregation of atmospheric, wastewater particulates, and solid waste.

Ultimately, any LCA study should offer a decision support to either decision or policy makers. While there are obvious limitations in some of the conclusions that may be derived such as determining whether energy consumption is more important than environmental emissions, however, a study that is vague and offers no direction to decision or policy makers will only compound the problem.

The work of Franklin Associates Ltd., compared the three diaper systems based on six criteria namely net energy requirements (using LCA method), net energy requirements (using closed thermodynamic balance), water volume requirements, atmospheric emissions, solid waste, and waterborne wastes. Johnson [1994] on the other hand, reported on the input requirements for cloth and paper diapers. Johnson notes that there is more chemical and water usage for cotton while softwood pulp had higher energy requirements. In this case study, we consider only the factors derived in the study by Franklin Associates Ltd., although Johnson's findings can be easily integrated in the framework of AHP. The analytic hierarchy network for this case is presented in Figure 6.4 below:

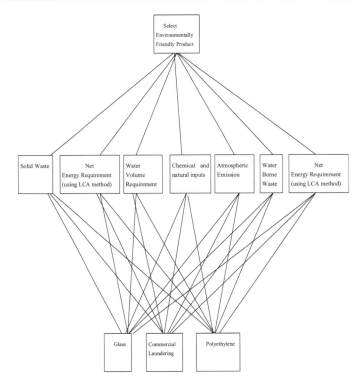

*Figure 6.4: An Analytical Hierarchy Network for Type of Diaper Selection Problem*

We use the information presented in the figures in LeVan's paper to generate a pairwise comparison ratio-scale for the six criteria and the three decision alternatives. These comparisons are shown in the Tables 6.2 to 6.7. Since the data we have in this problem is quantitative, rather than use weight assignments, we used ratios derived from the actual data to derive these tables. The values are then normalized using the method of AHP to obtain the priority indexes for each of the diaper type for any given environmental burden.

*Table 6.2: Pairwise Comparison of the Three Types of Diaper based on Solid Waste Generation*

| Type of diapering system | Single-use diapers | Commercial laundering | Home laundering |
|---|---|---|---|
| Single-use diapers | 1 | .48 | .40 |
| Commercial laundering | 2.083 | 1 | .833 |
| Home laundering | 2.5 | .1.2 | 1 |

*Table 6.3: Pairwise Comparison of Diaper Types based on Net Energy Requirements using LCA Methodology*

| Type of diapering system | Single-use diapers | Commercial laundering | Home laundering |
|---|---|---|---|
| Single-use diapers | 1 | 1.21 | 1.36 |
| Commercial laundering | .826 | 1 | 1.12 |
| Home laundering | .735 | .892 | 1 |

*Table 6.4: Pairwise Comparison of Diaper Types based on Water Volume Requirements*

| Type of diapering system | Single-use diapers | Commercial laundering | Home laundering |
|---|---|---|---|
| Single-use diapers | 1 | 2.25 | 2.813 |
| Commercial laundering | .444 | 1 | 1.25 |
| Home laundering | .355 | 8 | 1 |

*Table 6.5: Pairwise Comparison of Diaper Types based on Atmospheric Emissions*

| Type of diapering system | Single-use diapers | Commercial laundering | Home laundering |
|---|---|---|---|
| Single-use diapers | 1 | 1.04 | 1.8 |
| Commercial laundering | .962 | 1 | 1.731 |
| Home laundering | .556 | .578 | 1 |

*Table 6.6: Pairwise Comparison of Diaper Types based on Wastewater Particulates*

| Type of diapering system | Single-use diapers | Commercial laundering | Home laundering |
|---|---|---|---|
| Single-use diapers | 1 | 6.571 | 7.143 |
| Commercial laundering | .152 | 1 | 1.087 |
| Home laundering | .140 | .920 | 1 |

*Table 6.7: Pairwise Comparison of Diaper Types based on Net Energy Requirements using a Closed Thermodynamic Balance*

| Type of diapering system | Single-use diapers | Commercial laundering | Home laundering |
|---|---|---|---|
| Single-use diapers | 1 | .935 | 1 |
| Commercial laundering | 1.07 | 1 | 1.07 |
| Home laundering | 1 | .935 | 1 |

Applying the synthesization method of AHP, we derive the priority of each diaper type given the six environmental burdens. This result is presented in Table 6.8.

*Table 6.8: Priority Indexes for the Three Types of Diapering System for each of the Environmental Burdens*

| Type of diapering system | Solid waste | Net energy -LCA | Water volume usage | Atmo-spheric emission | Waste water particulates | Net energy - thermo |
|---|---|---|---|---|---|---|
| Single-usage diaper | 0.179 | 0.391 | 0.556 | 0.397 | 0.774 | 0.326 |
| Commercial laundering | 0.373 | 0.322 | 0.247 | 0.382 | 0.118 | 0.349 |
| Home laundering | 0.448 | 0.287 | 0.197 | 0.221 | 0.108 | 0.326 |

From Table 6.8, it could be observed that the preferred choice is Home laundering if the only concern is with solid waste disposal since it has the highest priority of 0.448. However, when the concern is with water volume usage, the preferred choice will be single-usage diaper. This is also the preferred choice when the concerns are with atmospheric emission and waste water particulates.

One problem with the application of the AHP in the Life cycle assessment study is to make a decision on which of the environmental burdens is more preferred to the other. As we noted earlier, such decisions may be affected by system boundary conditions since needs and requirements may vary from different geographical settings. Therefore, results derived may vary depending on where the life cycle assessment is to be applied. This will require us to compare all the six environmental burdens. This is a difficult task since none of the environmental burden should be considered an option. However, given the fact that we must reach a decision that will unavoidably contribute to environmental pollution; it is best to make choices that will lead to minimizing environmental pollution. It is recommended that a group of experts knowledgeable about this problem be used since weights assigned will affect the final recommendation. Table 6.9 is a presentation of the assignment we have made for the sake of illustration.

*Table 6.9: Pairwise Comparison of Environmental Burdens*

| Type of Environmental burden | Solid waste | Net energy - LCA | Water Volume Usage | Atmospheric Emission | Water-borne waste | Net energy-Thermo. |
|---|---|---|---|---|---|---|
| Solid waste | 1 | 2 | 4 | .25 | .2 | 2 |
| Net energy – LCA | .5 | 1 | 2 | .167 | .125 | 1 |
| Water Volume Usage | .25 | .5 | 1 | .143 | .111 | .5 |
| Atmospheric Emission | 4 | 6 | 7 | 1 | .5 | 3 |
| Waste water particulates | 5 | 8 | 9 | 2 | 1 | 4 |
| Net energy - Thermo. | .5 | 1 | 2 | .333 | .25 | 1 |

Based on the method of synthesization again, the following priorities were derived as shown in Table 6.10.

*Table 6.10: Priority Indices for Types of Environmental Burden*

| Type of environmental burden | Priority indexes |
|---|---|
| Solid waste | 0.114 |
| Net energy requirement - LCA Method | 0.061 |
| Water volume requirement | 0.037 |
| Atmospheric emission | 0.284 |
| Waste water particulates | 0.426 |
| Net energy requirement - Thermodynamic balance | 0.078 |

From the results of Table 6.10, it is seen that the three most important environmental burdens to consider in order of importance are waste water particulates, atmospheric emission, and solid waste. These results will have an effect in determining which of the different types of diaper system will be the preferred choice. We must also note that industrial policies must align with national policies on total environmental quality management. Priorities generated on environmental burdens should in fact, be in conjunction with policy makers and stakeholders from the different industrial sectors and environmental interest groups.

The final phase of this analysis will be to take the product of Table 6.8 and Table 6.10. Since Table 6.8 is a 3 x 6 matrix and Table 6.10 is a 6 x 1 matrix, the product of Table 6.8 x Table 6.10 will lead to a 3 x 1 matrix that will contain the priorities for the three types of diapering system. This result is presented in Table 6.11.

*Table 6.11: Priority Indexes for the Three Types of Diapering Systems*

| Types of diapering systems | Priority indexes |
|---|---|
| Single-usage diaper | 0.533 |
| Commercial laundering | 0.257 |
| Home laundering | 0.210 |

The result suggests that given the six environmental burdens and the comparisons provided by experts on them, the preferred choice should be to implement a single-usage diapering system. The next preferred choice is a commercial laundering system.

We must however, point out that this case study is for illustrative purposes only. There may be other critical factors such as the input factors identified by Johnson that were excluded in our analysis. Also, the use of expert judgment in assigning some of the weights introduces subjectivity in the model. The weight assignments may change for different situations and system boundaries. However, this approach provides a guide on how important decisions on LCA studies may be reached. In this chapter, we have excluded some information on the use of AHP such as computing consistency index and how to use AHP for group decision making. However, these issues need to be explored before applying AHP. Further reading on AHP can be obtained from the following references Madu [1994, 1999], Madu and Georgantzas [1991], Madu and Kuei [1995] and Saaty [1980, 1987].

## Conclusion

In this chapter, we have introduced the concept of life cycle assessment. We note that this requires product stewardship where the manufacturer takes a cradle-to-grave approach of its products. There are two main definitions of life cycle assessment that frequently appear in the literature. These two definitions provided by SETAC and ISO have also been discussed. Further, life cycle assessment consists of three main components: life cycle inventory analysis, life cycle impact analysis and life cycle improvement analysis. We have discussed all these and noted that life cycle inventory analysis is the foundation of all life cycle assessment. If that phase is incorrectly done, the entire process will be flawed. We have also identified some of the limitations and benefits of life cycle assessment and we have shown with a case study on types of diapering systems how the analytic hierarchy process (AHP) could be used in the context of life cycle assessment decision-making. We also discussed a strategic framework for LCA.

Another important area that was discussed in this chapter is life cycle cost assessment. This is an important issue since we depend on the co-operation of businesses that have profit motives to ensure that LCA is successful. As a result, we need to speak in a language business executives will understand, by exposing them to the cost of poor environmental quality systems.

We also noted that industrial environmental goals must support national policies and in fact, derive from national policies on environmental protection. The guidelines, targets, and priorities on environmental burden should be set together by businesses working together with government agencies. Finally, life cycle assessment can help guide decision-makers to produce environmentally friendly products. It sensitizes them about the needs of their stakeholders and by listening to the voices of the stakeholders, quality decisions on environmental issues that can help improve the bottom-line of the corporation and make it competitive can be derived.

## Reference

Fava, J.A., R. Denison, B. Jones, et al., eds., (1991) A Technical Framework for Life-Cycle Assessments, SETAC, Washington, D.C. 134 pp.

Franklin Associates, Ltd. (1992) Energy and environmental profile analysis of children's single use and cloth diapers, Franklin Associates, Ltd., Prairie Village, Kan. 114 pp.

Johnson, B.W., (1994) "Inventory of Land Management Inputs for Producing Absorbent Fiber for diapers: A comparison of cotton and softwood land management," Forest Prod. J. 44(6): 39-45.

Kirkpatrick, N., "Life Cycle Assessment," Retrieved 12/21/99 from http://www.wrfound.org.uk/previous/WB47-LCA.html.

LeVan, S. L., (1995) "Life Cycle Assessment: Measuring Environmental Impact," Presented at the 49th Annual Meeting of the Forest Products Society, Portland, Oregon, June, pp. 7-16.

Madu C. N., (1999) "A Decision Support Framework for Environmental Planning in Developing Countries," *Journal of Environmental Planning and Management*, 42 (3): 287-313.

Madu, C.N. (1996) Managing Green Technologies for Global Competitiveness, Quorum Books, Westport, CT.

Madu, C.N., (1994) "A quality confidence procedure for GDSS application in multicriteria decision analysis," *IIE Transactions*, 26 (3): 31-39.

Madu, C.N., & Georgantzas, N.C., (1991) "Strategic thrust of manufacturing decisions: a conceptual framework," *IIE Transactions*, 23 (2): 138-148.

Madu, C.N., (2006) House of Quality (QFD) in a Minute, 2$^{nd}$ edition, Fairfield, CT: Chi Publishers.

Madu, C.N. & Kuei, C-H., (1995) "Stability analyses of group decision making," *Computers & Industrial Engineering*, 28(4): 881-892.

Madu, C.N., (1998) "Strategic Total Quality Management," in Handbook of Total Quality Management, (ed. C.N. Madu), Boston, MA: Kluwer Academic Publishers, pp 165-212.

Product Ecology Consultants, "Life Cycle Assessment (LCA) Explained," Retrieved from http://www.pre.nl/lca.html on 12/21/99.

Saaty, T.L., (1980) The Analytic Hierarchy Process, New York, NY: McGraw-Hill.

Saaty, T.L. (1987) "Rank generation, preservation, and reversal in the analytic hierarchy decision process, Decision Sciences, 18: 157-162.

SETAC (North American & Europe) Workgroups, "Evolution and Development of the Conceptual Framework and Methodology of Life-cycle Impact Assessment," SETAC Press, Washington, D.C., 1-14.

Standards Council of Canada (1997) "What will be the ISO 14000 series of international standards - ISO 14000," January.

## Chapter 7

# Design for the Environment – Part I

The growing concern about the need to protect the environment and curtail the rate of consumption of earth's limited and nonrenewable resources has pushed manufactures and service providers to design their products and services for the environment. Designing for the environment has become a critical component in achieving competitiveness. "Design for the Environment" (DfE) is a means of incorporating environmental component into products and services during the initial product design stage. The aim is multifold and includes optimal utilization of limited resources and material, waste reduction, improved designs, reduced product liability, and improved efficiencies. Products and services that are designed for the environment are environmentally responsible, satisfy the needs of the customer, and help the company to be competitive and improve its market positioning.

Design for the environment understands the need to be environmentally conscious but also recognizes the importance of including all the quality attributes of importance to the customer in the design stage. DfE is a product innovation strategy that seeks ways to minimize cost and improve performance by examining the product through its life cycle. Product assessment starts from the extraction point through its end of life where components may be disassembled, reclaimed and reused or disposed as waste. The strategy is to find the most feasible and effective ways to do all these without harming the natural environment.

Design for the environment is a strategic response to achieving sustainable production. It relies on several techniques to identify environmental impact and improve performance of the product through its life cycle. The major emphases in this design approach are on:

- Design for recycling
- Design for energy efficiency
- Design for remanufacture
- Design for disassembly

- Design for disposability
- Design to minimize hazardous material

There are several benefits to the Design for the Environment. We shall expand the discussion on some of the benefits identified by Yarwood and Eagan.

## Reduced Cycle Time

Time-to-market is critical in today's market environment. Product life span is very short and there is increased proliferation of new products. In a highly competitive environment, it is essential that manufacturers and service providers adopt a rapid response strategy in satisfying the needs of the market place. Inability to respond timely to the market with products that meet environmental quality amongst all attributes may lead to loss of market share. It is important that the cycle time to new product introduction is reduced. This can be achieved when environmental concerns are integrated in the design process to ensure that the product does not face either recall or rejection from the marketplace.

## Reduced Costs

Top management reacts well to costs. In fact, one way to get top management to adopt ideas related to environmental quality is to expose them to the cost of poor environmental quality. There are many types of costs that can be borne out of poor environmental quality. More notable among these are the costs of disposing waste, cost of cleaning up environmental waste, cost of reworking, scraps, and rejects, and the liability costs and fines that may be associated with not meeting environmental regulations. However, the greatest cost of all is the loss of customer goodwill. In order for a company to be competitive, it must cut down on cost. Consequently, it must improve its bottom-line. Designing for the environment enables a manufacturer to not only cut on all these costs but also to improve its bottom line and position itself to gain market share

## Improved Products

Designing for the Environment can also be viewed as a quality strategy. It improves the quality of the product by critically assessing all the environmental attributes at the design stage and developing and evaluating alternative substitutes. In other words, the product composition is made up of attributes or components that would satisfy the needs and wants of the customer. Quality is often defined as customer satisfaction. What pleases or delights the customer is customer satisfaction. By responding to the needs of the customer to improve the environmental content of the product, an improved product is designed and developed. With such products as "delighters," the manufacturer can continue to expect the patronage and support of its customers.

## Reduced Regulatory Concerns

Pollution prevention pays. There is an avalanche of regulations on environmental protection. A competitive company cannot be reactive but rather proactive and identify and prepare itself to future trends in environmental compliance and regulation. Proactive companies can develop strategies to adapt them to these future trends by designing their products and processes to respond to environmental needs. Companies of the future go beyond today's information to anticipate future possibilities. They seek out the most environmentally conscious and feasible methods to design and produce their goods and services knowing fully well that they are accountable to the role of their products in the natural environment. Regulatory agencies and interest groups are on high alert at the customer's behest and can anticipate potential dangers to the environment. Many countries and world bodies such as the United Nations as well as local, state and federal agencies are increasingly enforcing new environmental laws and demanding that manufacturers change and abandon high risk environmental practices. Now, the current trend for companies is to place environmental certification labels on their products as a mark of environmental accomplishment. Many countries have also adopted ISO 14000 standards and insist that manufacturers adopt such standards. The recent ratification of the Kyoto Protocol by Russia's lower house of parliament shows that indeed the enforcement of reduction on greenhouse gases is around the corner. Such a reduction will more likely lead to introduction of newer products and processes that are more efficient. The time-to-market becomes very critical to reactive companies that never anticipated a major world power to support the Kyoto Protocol.

### Reduced Future Liability

This can be better explained using the 3P concept – Pollution Prevention Pays. There is no statue of limitation when it comes to environmental pollution. Polluters are held responsible any time the impact of their activities on the natural environment or natural resources are detected. It is up to the manufacturer to design its products and services to have less impact on the environment and to have less demand on limited natural resources. When companies design environmentally friendly products that are sustainable, they are able to reduce the risks associated with environmental liabilities and the costs associated with them.

Environmental liability laws are already in place in many countries. For example in the U.S., there are several federal statues that require financial assurance for environmental liabilities. Prominent among these statues are the Oil Pollution Act (OPA), the Resources Conservation and Recovery Act (RCRA), and the Comprehensive Environmental Response, Compensation, and Liability Act (CERCLA). These acts take a cradle-to-grave approach and hold polluters responsible for damage to natural resources. Polluters pay financial liabilities, which may include clean up costs, real and personal property damages, lost government revenues, and other related costs. Design stage offers the ability to manufacturers to review their design strategies, product components, requirements, and emissions, and develop more environmentally sensitive product.

## Improved Market Position

Companies compete by identifying a particular market niche they can satisfy. The "green" market is booming and in vogue today. More and more consumers are increasingly health and environmentally conscious. They seek products that would not harm them or the environment. Environmental quality has therefore taken a new role in the marketplace. Successful companies are able to listen to their customers, understand their needs and wants and design such into their products. Those companies that have been able to do so continue to reap the benefit and advertise widely their environmental efforts as part of their social responsibility function to the society. For example, manufacturers of appliances such as washing machines, dishwashers, dryers are increasingly using the "energy star" label on their products to show that these products are energy efficient. They are not only beneficial to the customer in terms of savings from energy cost but they are also environmentally friendly by having lesser demand on the non-renewable energy resources. The same is also true with auto manufacturers who are increasingly designing lighter vehicles that are fuel-efficient.

## Improved Environmental Performance

Achieving environmental quality is a global issue. It transcends beyond national boundaries. The world at large has recognized the significance and that is why international bodies such as the United Nations have taken the lead to discuss some of the common environmental issues that face the world as a whole. The growing concern on toxic and waste management, global warming, and depletion of the ozone layer impact the earth as a whole and do not reside in any one country. A recent article[14]warned about human's reliance on the consumption of nonrenewable resources and the subsequent destruction of the Earth's ability to sustain life. The World Wildlife Fund noted in its regular Living Planet Report, that the United Arab Emirates, the United States, Kuwait, Australia and Sweden are the biggest consumers of nonrenewable natural resources. It is estimated that humans currently consume 20 percent more natural resources than the Earth can produce. All these concerns further highlight the importance of integrating environmental issues into the design process in order to achieve superior environmental performance. This also shows why the world was rejoicing when in October 2004, the Russia's lower house of parliament ratified the Kyoto Protocol. Even though the treaty was rejected by the United States, Russia's ratification marks the final acceptance needed for the treaty to pass. Other industrial nations such as Japan and Germany have also adopted the treaty.

## Social Responsibility Function

Businesses today are increasingly aware of the need to develop sustainable production systems. They have responsibility of protecting their environment, employees and stakeholders and therefore, need to design for the environment. Corporate policies are not made entirely based on bottom-line but are also influenced by the community's perception of corporate operations and agenda. Good corporate image must be maintained. The business must be viewed in positive light by its stakeholders for it to sustain its operations. It is therefore imperative that corporations respond to the call to develop environmentally conscious processes and products. Responding to the stakeholders' environmental needs is good business and may often stimulate radical changes in the

---

[14] Fowler, J., "Group warns on consumption of resources," Associated Press, October 22, 2004,
http://news.yahoo.com/news?tmpl=story&cid=624&u=/ap/20041022/ap_on_sc/ plundered...

ways businesses are done. Processes and products may need to be completely redesigned and attitudes may need to be changed. Organizational mission, vision and mode of operation would all be affected. Environmentally consciousness is a strategic issue that businesses cannot overlook. Worldwide policies dictate environmental changes as well as market demand and competition and companies must react to remain competitive. Environmental management is a social responsibility function, which business organizations must provide to their business community and society as a whole. It is the key to competing in the 21$^{st}$ century.

## Design Strategies

Design for environment (D$f$E) is a systematic and systemic consideration of environmental factors in product and process design. In order to achieve this goal, D$f$E must take a holistic view of the product by studying the effects of the design through the product's life. This product lifecycle approach requires that the interaction of the product with its natural environment be explored and attempts made through the design to ensure that environmentally sensible or conscious products are designed. We shall develop strategies in this section on how design for environment can be achieved. Some of the strategies developed here will benefit from our knowledge of tools and techniques for quality management.

## Design Team

The first step in designing for environment will be to form a design team. The design team will be a team that is made up of important stakeholders. These stakeholders are those whose reactions and actions affect or are affected by the product. The team is made up of people with diverse and interdisciplinary backgrounds and different worldviews. They could include members of important interest groups. The role of the team is to identify potential environmental impacts of the product design.

They review alternative product designs and identify associated environmental problems and benefits. Such may include analysis of the material requirement, disposable issues, distribution and logistics, and recycling considerations. The team in a sense will conduct some form of product lifecycle assessment. The team works with design engineers to address *"what if"* or sensitivity analysis, review alternative designs and scenarios, brainstorm and generate ideas on how some of the environmental issues may be resolved. This team engages in optimization with the goal of identifying the best product design to make efficient use of natural resources in a systemic perspective. However, product life assessment is not easy to achieve. It is difficult to quantify benefits and sometimes what may be thought as benefit may actually create more environmental hazards. Therefore, the goal should be to make the best judgment given all the information available.

The team must be empowered. It must have a decision-making ability and should have the support of the top management. The team reviews the different scenarios or alternatives and assigns priorities to those scenarios. Thus, a framework can be derived of "stakeholder" requirements and matched against "design requirements." By prioritizing stakeholder requirements, the appropriate design strategies can be identified. We shall lay a framework that could be useful for design for environment.

## Stakeholder Requirements

First and foremost in designing for environment is to understand stakeholder requirements. What is it that the stakeholder wants? The wants and needs of the stakeholder must be designed into the product or service to delight the stakeholder and get its acceptance of the product or service. The definition of stakeholder is inclusive of customers as well as all those that can affect or are affected by the product or service. Therefore the intrinsic needs of the customer are also considered while designing the product but at the same time, the impact of the product on its external environment is considered to achieve a tradeoff. How then do we identify these needs?

There are several approaches that may be adopted to fully understand the wants and needs of the stakeholders. Some of these approaches may be borrowed from the quality management literature where there is a great emphasis on producing products to satisfy the needs and wants of the customer. But rather than restricting this to the needs and wants of the customer, we expand it to the needs and wants of the stakeholder. Madu [2006] identified how customer needs and wants are solicited. Some of them are discussed below:

## Environmental Quality Dimensions

Madu [2006] referred to customer requirements as synonymous to "quality dimensions." In this book, stakeholder requirements are what we refer to as environmental quality dimensions. They represent the attributes or features of a product or service that the stakeholder demands in order to achieve environmental quality. These attributes may be perceived differently and may also have different importance weighting to the stakeholder. For example, the need to reduce the energy consumption for a dishwasher may or may not have higher priority to the stakeholder when compared to the water consumption for a wash. Simply stated, all attributes are not the same. Some are more important depending on their impacts on the environment. However, it is not easy to identify which attributes are more important than the other in an environmental context. A thorough analysis would require an environmental impact assessment, which is difficult to conduct.

The use of stakeholder teams in identifying environmental quality attributes may lead to a long list of attributes that have to be grouped and rated in terms of importance. It may be necessary to develop an "environmental assessment survey" where experts may be used to develop some priorities on identified environmental impacts. This way, there is a focus in the design phase on which environmental impacts to focus on.

Critical incident approach could be used to develop the environmental assessment survey. This approach is based on analyzing the product's environmental impact by asking the stakeholder to identify how environmental attributes of a product or service may positively or negatively affect its perceived organizational performance. The aim is to identify the positive impacts that influence stakeholders' perceptions of the product or service and to reduce or develop alternatives to the negative impacts.

In designing for environment, the key is to look at alternative design strategies and select the option that would have minimal environmental impact. Such selection strategies should be able to analyze carefully, information obtained from stakeholders. Stakeholder analysis may be based on surveys that address the following issues:

## *Product Content Environmental Questionnaire*

What is the recyclable content of the product?

Could the recyclable content of the product be increased? And by how much?

Are there alternative renewable materials that could be used?

What is the environmental content of all the material used in product development?

Are there more environmentally friendly materials to replace some of the existing materials?

How much energy is needed in the manufacturing process per unit of the product?

How much energy does the process expend?

How much waste is created in the process?

What is the total energy cost including transportation costs?

How much waste in material and energy are created along the supply chain?

How would you rate the ease of recycle of the waste product?

How would you rate the ease of disassembly and material recovery?

How much solid waste is created at the end of products life?

What type of emission is generated from the production process?

What type of emission is generated from the product?

Is the emission recoverable and usable?

How much hazardous wastes are produced?

Is the waste oil recyclable?

Is the wastewater recyclable or treatable?

Are any of the by-products or liquid wastes treatable or reusable?

Are any of the gases generated among the banned or prohibited gases?

Are any of the gases generated among the greenhouse effect gases?

What materials are used in packaging the product?

Are the packaging materials recyclable?

Can the product be easily disassembled, repaired or reused?

Are there any product components that cannot be reused or recycled at the end of the product life?

This product content environmental questionnaire identifies some key questions about the product, process, and distribution. It could be expanded and modified. However, the key behind its use is to be able to compare alternative product designs based on their environmental contents. These questions can be grouped into areas such as energy usage, product content, end of life etc., if it makes it easier for analysis. However, it addresses issues that stakeholders would want to investigate before selecting one design strategy against the other. The different design strategies can be compared on this basis with the goal of finding and selecting the one that has the least environmental burden. Again due to the fact that there is a long chain involved in any production and distribution system and it is difficult to say which environmental burden is more desirable, it is almost impossible to get an *optimal* design that will guarantee the least environmental burden. When we assess a product design, we take a cradle-to-grave approach by studying the product through its lifecycle. That entails tracing the product from the extraction phase through design and production to distribution and logistics and then to consumption and through its end of life and disassembly and disposal. It is indeed a difficult task to optimize this process.

## Pareto Chart

One of the popular methods to organize errors, problems, or defects found in a process or product is known as the Pareto chart. An economist Vilfredo Pareto developed Pareto chart in the nineteenth-century but its use in the management field was made popular by one of the leading founders of quality management Dr. Joseph M. Juran. According to Juran, 80% of a firm's problems are a result of only 20% of the causes. This concept has worked very well. In the present context, Pareto chart could be used to identify the 20% of the causes that lead to 80% of the environmental burden we observe in the process or product. By focusing on these "critical" causes, we can eliminate most of the environmental burdens. This way, solving the critical problems that explain most of the environmental burden the firm is able to make efficient use of organizational resources. An illustrative diagram of the Pareto chart is presented below as Figure 7.1. The data used to construct the Pareto chart is derived from Table 6.10 and represents the frequency distribution of the diapering system. Using the Pareto chart, it shows that waste water particulates and atmospheric emission consist of most of the problems created by the diapering system. These two environmental burdens make up 71% of the problem. The idea is to focus on the critical few in solving a problem. Therefore, to address the environmental burden as a result of diapering system, the major focus should be on these two problems.

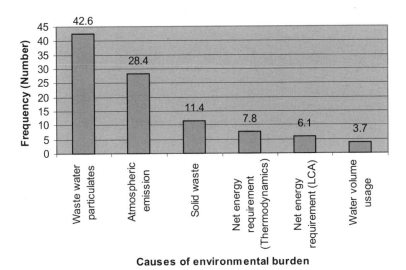

*Figure 7.1: Pareto Analysis of Environmental Burden from Diapering System*

## Ishikawa Diagram

Ishikawa or fishbone diagram is a cause-and-effect diagram and could be used to identify the major causes of environmental burden in a product or process. This problem solving technique is based on the premise that there are four major causes to any effect and these are known as the 4 *M*s – Man, Material, Machine and Methods. This method of identifying environmental burdens was discussed in more details in Chapters 5 and 6.

## Analytic Hierarchy Process (AHP)

Analytical Hierarchy Process is also a problem solving technique that could be used to evaluate alternative strategies. In designing for environment, it is important to review different design strategies and identify the ones that may create less environmental burdens. Due to the problems associated with life cycle assessment in the sense that it is often difficult to quantify environmental impacts or to suggest that one environmental impact is more preferable to the other, it would be difficult to generate an "optimal" design. The AHP is a multi-criteria decision making model that encourages the participation of the decision or policy maker(s) in assigning preferences to the different design options or alternatives. It is possible that the design strategies adopted may be different for different environmental settings. For example, a community that has severe water shortages but has abundance of energy supply such as solar energy may find a design option that demands significant water consumption and usage as unattractive when compared to the need for energy consumption. The process of AHP is discussed in Chapter 6 where it is applied in life cycle assessment problem.

The key question here is how the AHP would help in designing for the environment. The role of the AHP would be to evaluate the different production alternatives given the several criteria and sub-criteria that may influence the organization's ability to produce an environmentally friendly product or services. AHP is broken down into three major levels which are discussed below.

# Goal

The goal outlines the objectives of the corporation in this case. The major objective here is to design and produce the best environmentally friendly product.

# Criteria

The goal cannot be made in isolation. There are several factors that would influence the attainment of this goal. These factors include the environmental burden that may be involved in the process, the financial and technical constraints. The aim here is not just to look at the environmental burden but also to understand the technical and financial feasibility of the process. However, we are going to focus on identifying the environmental burdens that may be identified at this stage. They are listed below:

- Dematerialization – The idea here is that the less the material requirement the better. This will imply lesser dependence on physical products and its implications in achieving sustainable production practices is obvious. When lesser physical material is used, there will be less energy, transportation, and consumption and disposal problems. Cost is also saved. This is accomplished by cutting down on the size and weight of products. Auto manufacturers are increasingly reducing the size and the weight of their vehicles; Cellular phones are getting smaller and smaller and some are pocket sizes; Computer hard drives are even smaller with the use of USB drives that have much higher memory and are yet much smaller in size and very light; Laptop Computers have also declined in both size and weight from where they started. One of the greatest developments in this area is the use of the Internet. Many of us today, do our shopping online, use online libraries and even attend online schools or take classes online. We all use email and have cut down significantly on the demand for snail mail. We subscribe to online newspapers and magazines and have cut down on the demand for physical newspaper delivery. All these activities have significantly cut down on the demand for physical products and the associated material, energy and distribution needed to produce these products, deliver, consume, dispose and recycle them. When we design for the environment, we need to be able to compare design strategies on the ability to dematerialize because this is one way we can truly design for the environment.

- Multiple Users – Multiple users for a resource could cut down on duplication of the same resource. This could be observed mostly with equipment that is commonly shared as in universities, hospitals or other organizations that provide services. This also has a lot of implications not just in duplicating the equipment themselves but also in cutting down material requirements. For example, suppliers often share information that are contained in a central database system such as ERP (Enterprise Resource Planning) system and such information are often obtained through the Internet. Suppliers with the right access can coordinate activities with the manufacturer without the traditional paper order processing. Such electronic activities can help to reduce demand for material and make the system more efficient.

- Multiple Function Capability – Can the product do more than one thing? If so, it is attractive and yet will demand lesser material. We have witnessed the evolution of telephones that are all integrated containing not just the regular telephone but fax, caller ID, phone directory, scanner, copier and printer. Yet, they are relatively compact compared to when each of these functions was represented by separate equipment. Products with multiple function capability may be less demanding on the environment.

- Value engineering – This involves component analysis. It is important that every component in a product add value to the product. Components that do not add value to the product need to be removed in the product design stage. A critical analysis of these components should be done to identify the primary and secondary functions and to see which may need to be removed without compromising the quality of the product.

- Reliability and Durability – These attributes are often overlooked but have significant influence on the demand for material, energy, transportation, and wastes. Undoubtedly, unreliable products would consume more materials and energy and create more scrap. Madu [2004] demonstrates the relationship between reliability and maintainability and sustainable development. He also discussed different methods that could be used to improve the reliability and durability of products. Durability while used together with reliability here, deals more with the robustness of the product. It will address the ability of the product not only to satisfy customer's expectations but also to withstand adverse environmental conditions. Products that are able to meet the reliability and durability conditions are also more likely to be of high quality.

- Ease of Maintenance and Repair – When products are easily maintained and repaired, the chances of error are reduced and there will be lesser need to transport the product back to the manufacturer for maintenance work that could have been easily done by the end user. This is common with computers and electronic equipment where users normally call the manufacturer's HELP line for technical assistance or log on to the Internet for Online Help. With easier instructions, the user can cut down on the need to send the product back for such services.

- Modular Product Structure – Many products are now designed to use modular components. This makes the product adaptable to new technological changes. When a new feature is introduced, the product is upgraded thus reducing the need to discard the product and buy a new one. For example, many personal computers have open slots to allow for upgrades. New memory chips or hardware components could be added over time without having to replace the entire computer system.

- Emissions – It is important to know the chemical composition of products. Some products may use chemicals that generate hazardous emissions during their lifecycles. Some of these emissions may be toxic and may include some banned elements or those that might affect the ozone layer. It is important that conformance to national and international laws be followed through a process of de-selecting hazardous or banned substances. Such should be designed out of the product.

- Renewable Materials – Efforts should be made to ensure that raw materials used are renewable and can be regenerated in a reasonable time. The use of non-renewable resources such as fossil fuels is not environmentally sound. Renewable materials are usually biodegradable and create lesser environmental burden than nonrenewable materials. However, choices have to be made between different forms of renewable materials to see which creates the lesser environmental burden over its lifecycle.
- Energy Requirement – It is important to understand the energy requirements of the product and process.

## Decision Alternatives

This deals with the decision choices or options that will present the lesser environmental burden.

## References

Yarwood, J.M., and Eagan, P.D., "Design for the Environment – A Competitive Edge for the Future ToolKit," Minnesota Office of Environment Assistance, Minnesota Technical Assistance Program (MnTAP), pg 1-16, undated.

Madu, C.N., 2006 House of Quality (QFD) in a Minute, 2nd edition, Fairfield, CT: Chi Publishers.

Madu, C.N., "Strategic value of reliability and maintainability management," *International Journal of Quality and Reliability Management*, (2004).

# Chapter 8

# Design for the Environment – Part II

In this chapter, we discuss how design for the environment could be achieved. We build on the previous discussion (Part I) by developing strategies and models to enable the design and production of green products and processes. In part I we mentioned the importance of stakeholder participation. We shall advance that discussion by using a business strategy that would enable the optimal utilization of resources in product design and development process. This business strategy is known as *concurrent engineering*.

## Concurrent Engineering

Concurrent engineering is a strategy by an integrated team to ensure that tasks are done in parallel and that all aspects of product development are considered in the design, development, production and distribution stages. By accomplishing tasks concurrently, the cycle time to introduce new product or service is reduced. New products or services are designed for produceability to reduce time to market. This strategy achieves a better match between the product and the process, enhances quality, and reduces waste and cost. The concurrency of the design function ensures rapid market response and gives the firm competitive advantage

Concurrent engineering requires new thinking in the organization. There is need for change management and for people and processes to adapt and continuously seek improvement. The integrative nature of the team requires a cross-functional view of the organization, the product, and the processes and such views must be adopted in product development to understand the lifecycle of the product. Knowledge of the product's lifecycle will facilitate the development of clear and concise environmental performance targets for both the short-term and the long-term. It also helps the team develop preferences, scenarios and evaluate alternative options based on the products or the process's environmental impacts or burden.

Through concurrent engineering, a holistic view of the product or process is taken by creating a *cradle to grave* approach as the cross-functional team evaluates how the product or process interacts with its extended environment through its lifecycle. One of the major benefits of concurrent engineering is in performance improvement. This strategy helps eliminate redundancy in the design and production process, the number of revisions, and prototypes. Effectively, it reduces the product cycle time and leads to designing the product right the first time. Thus, wastes and scraps that result from repetitive operations or reworking defects can be avoided and energy and resources are conserved.

Concurrent engineering is a shift from the traditional manufacturing process. It follows a horizontal process that is lateral and flat rather than the hierarchical vertical process that is implemented in the traditional approach. We shall compare both the vertical and horizontal product realization processes to show how this new business strategy could play a major role in designing for the environment.

## Vertical vs. Horizontal Product or Process Realization

In the traditional vertical product realization process, marketing information is the key to new product development or redesigning of existing products. It is presumed that customer information, needs and wants are obtained by the marketing department. There is no integrative team and the functional departments in the firm are all independent. Information obtained by marketing is the motivation for either new product development or redesigning of an existing product. Such information is then passed over to manufacturing whose role is to design and produce the product as they see fit. Marketing may further assist through market surveys to identify key features that should be in the product design. The hierarchical process adopted by vertical product realization could lead to long cycle times and may not effectively capture the wants and needs of the customer thereby leading to waste generation. Figure 8.1 shows this hierarchical process.

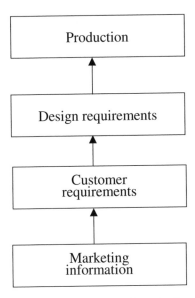

*Figure 8.1: Traditional Manufacturing Process*

Some of the problems with the vertical product or process realization approach are outlined below:

- The vertical product realization approach is based on functional silos that are independent. The different departments do not operate in unison and there is no cross-functional and integrative team. A sequential task procedure is involved where different roles are differentiated and not worked in parallel. For example, marketing, design, production conceptualization, manufacturing and distribution are all treated independently and are not looked at to see if there are interdependence and duplicities that may be present. The resulting effect is that certain tasks may be repeated several times, wastes may be created, and it will take longer time to produce and deliver the product. Higher environmental burden is created as these efforts consume more resources and energy and in the long run, lead to more waste creation. When tasks are done in parallel, there is transparency and it is easy to identify avenues to cut down waste so resources can be efficiently utilized and optimized.

- Duplication of tasks and waste create environmental burden that need to be checked from the birth of the product concept. When designing is separated from production and distribution, it is possible that the finished product may not be feasible and even when produced; it may not be environmentally friendly. When the entire process of production is considered at the early stages by checking the designability, manufacturability and disposability of the product, it becomes easier to create an environmentally friendly product. No aspect of product management should be treated in isolation. The design team should work with production and distribution teams to form an integrative team. A complete picture of the product should be developed by looking at its entire lifecycle. Designers should not see distribution issue as separate from their function but as part of their goal to design an optimal product that will minimize environmental burden.

- Quality is compromised. The intrinsic and extrinsic needs of the customer must be satisfied to achieve quality. Customers today are concerned about environmental impacts of the product. They want to ensure that the product they support does not create an environmental burden that could ultimately affect their quality of life. Also, customers understand the need to conserve limited natural resources. When design, production, or distribution is separated, there is tendency of reworking and creating of scraps and rejects when one stage of the production process is found to be incompatible with any other stage. Such waste creation becomes an environmental burden that could have been avoided if all the stages of production were worked in parallel.

- The vertical product realization process leads to longer lead time to introduce new products. Aside from the fact that this long lead time makes it difficult for the firm to respond rapidly to market changes and thereby losing its competitiveness, it creates waste. Products today have shorter lifecycles. When a firm fails to respond to the changing needs of the customer, it may be stocked with obsolete products that may be disposed as waste. Such wastes have chain reaction effects on the environment. They have effects in terms of solid waste disposal, energy that is wasted in producing and distributing them, and materials and resources that were consumed. Long lead time to introduce a product is a potential source of environmental burden.

- It is important to manufacture the product correctly the first time. Failure to accomplish that would lead to waste creation. Companies that are competitive have higher productivity. Creation of waste adds cost both in terms of environmental clean up, fines and punitive damages but also costs the company customer good will. When items are produced right the first time, the cost of production is lower and the company can achieve customer satisfaction.

To overcome the problems listed above for vertical product realization, designing for environment requires a new focus. The focus should be on a structure that encourages rapid information sharing and integrative team formation without the hassle of hierarchical structure on information flow. This flat structure for the product realization process is known as *horizontal product realization process* and it is shown in Figure 8.2. We shall discuss this process and how it could help to design for the environment.

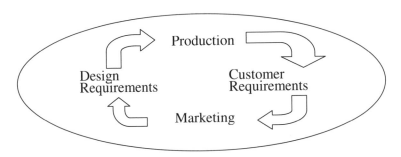

*Figure 8.2: Concurrent Engineering*

## Horizontal Product Realization Process

This approach involves lateral flow of information between the major product design and development process. As shown in Figure 8.2, customer, production, marketing and design have a single view of information and work in parallel to achieve the rapid development of products. Information is obtained and analyzed at the same time. The members of the cross-function department work in unison in product design and development. Customer requirements are readily matched with design requirements with the different functional units working in parallel to find the best way to design, produce and distribute the product to meet both the intrinsic and extrinsic needs of the customer. This single view of the product design helps to produce the right product and to do it right the first time. The different functional units working together come to understand and appreciate each others' role and work together to achieve sustainable products.

Horizontal product realization process can help to address some of the problems identified with the vertical product realization process. Already, the flat structure of this process ensures rapid and lateral flow of information that could be shared by all participants and used to evaluate alternative design requirements to match customer requirements. Therefore, *green* products that meet environmental constraints can be developed, produced and distributed. As we look at the value chain that is achieved through this process, we notice subtle gains in the product development process that are important in achieving sustainable or environmentally conscious manufacturing. We look at some of these below:

- Product design and development – Product design evaluates sustainable issues such as design for recyclability, design for disposal, design for disassemble, design for remanufacturing, and design for manufacturability. Each of these design strategies consider the natural environment to ensure that sustainable products are developed. Sustainable products are environmentally friendly and would have minimal environmental burden. With the use of cross-functional teams, different design strategies are evaluated and lifecycle assessment as well as environmental impact assessment is conducted to identify the appropriate design strategy.
- Quality is still the key. When products are optimally designed, they meet their intended purposes without harming the environment. They

conserve natural resources and they emit less environmental burden. Poor quality leads to higher environmental cost. Hence the adage that prevention pays. Concurrent engineering helps to adopt a focus on quality that considers the product's impact on the natural environment. The use of concurrent engineering improves quality and productivity and increases the quality of life of the workers. It helps to reduce the cost of production and the delivery and introduction time of the product to the market. Efficiency is achieved and products that meet the needs of the customers and stakeholders are designed and produced.

- Competitive businesses adapt and respond well to their markets. When the right products are developed, the business gains market share, becomes competitive, and can innovate. Green products are vogue and there are challenges that businesses face. Innovative organizations need to face these challenges and continue to design and produce products that will continue to meet the needs of the customer.

## Quality Function Deployment (QFD)

The aim of QFD is to listen to the voice of the customer and systematically translate customer's requirements through each stage of the production planning and development to requirements that the product must meet. Therefore, it matches customer requirements with design requirements so that *quality* can be designed into the product. Our consideration of quality here goes to include both the intrinsic and extrinsic attributes of the product. In other words, a quality product is viewed not only by the product's performance for its intended usage but also on how the product through its lifecycle, interacts with the natural environment. A quality product in this context should be able to meet stakeholders' needs and react favorably to the natural environment. QFD has been widely applied in the context of environmental management. Some have developed a multi-stage process to QFD referred to as GQFD (Green Quality Function Deployment). GQFD involves two phases. In Phase I, three houses for quality, environment, and cost are developed and used to compare product concepts. The indexes that are obtained for the three houses are then fed into a hierarchical structural framework in Phase II where the analytic hierarchy process (AHP) is used to select the *"best"* product concept [Bovea and Wang]. Of course, the best product concept may never be found by this approach. There are several factors or criteria that are considered in determining the best product concept and some are intangibles that are very subjective and difficult to quantify. The perceptions of the stakeholders who participate in comparing these concepts may very well influence the choices that are made. Furthermore, lifecycle assessment is very difficult to make. It is often difficult to clearly document the environmental burdens and chain reactions involved in any particular process or product development. Even when that is done, it is difficult to clearly estimate the impacts of different environmental burdens to state which environmental burden is more important than another. In an area that is heavily afflicted with drought, water conservation may be more important than disposal of waste to landfills. This is not to say that one environmental burden is more important than the other however, there are situational circumstances that may make one more preferable.

Application of QFD should be integrated in the product planning and development. The goal should be to *design* green products. QFD should be used for effective planning and life cycle assessment issues should be duly integrated in the design process. We shall illustrate this process using a case study of paper production or paper recycling [Madu et al, 2002].


## Green Design Framework


The customer is the essence of any business. Businesses must design and manufacture products and provide services that meet customers' needs and expectations. It is, therefore, important that the customer is part of the products and services delivery process. Most literature on environmental quality management systems start by noting that there has been a growing interest in making products and providing services that are environmentally friendly. This increased interest in environmentally friendly products and services is, to a large extent, due to growing concerns about environmental pollution and disturbances by the general public. People are concerned about numerous environmental challenges, such as the quality of the air, the lack of green space in their communities, groundwater supply, the greenhouse gases and the damages to the ozone layer. They understand that the inability to manage these environmental problems will, ultimately, degrade the quality of life on earth. As a result of these concerns, many nations and international agencies have embarked on developing new guidelines and legislatures that can help manufacturers as well as consumers to be more environmentally conscious. For example, many nations now have the equivalent of the United States Environmental Protection Agency, which as the name shows, operates by developing and guiding policies to protect the environment. Also, international bodies such as the United Nations and the International Organization for Standards (ISO) have been very active in developing guidelines and procedures to effectively manage environmental pollution. Understanding the role of consumers will help manufacturers and service providers to design products that meet consumers' expectations and will, in turn, give them a competitive edge. It is important that manufacturers start their design from the viewpoint of the consumer or major environmental stakeholders. For example, the framework shown in Figure 8.3 starts with a particular objective. In the paper-recycling problem, the customer needs to select a recycled paper that meets certain standard needs. The quality or the grade of paper selected depends on the specifications of the print job the customer will undertake. For example, for jobs that demand high quality, recycled white waste paper may be of utmost importance; low quality recycled papers such as old newspapers may be sufficient for other jobs. Consumers can rate their application orientation by basing quality in the following order:

- White waste paper.
- Packaging waste paper or brown paper.
- Households or 'mixed' waste paper.
- Coated paper.
- Old newspaper.

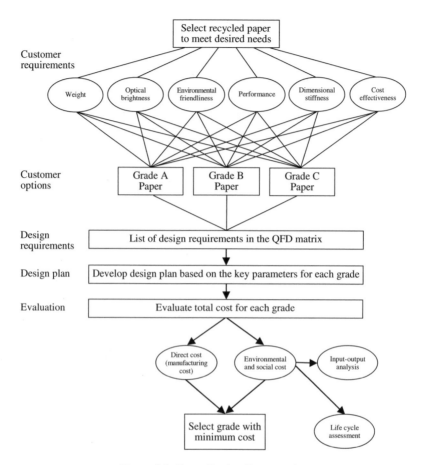

*Figure 8.3: Green Design Framework*

Once the decision is made on the application, a determination of the customer requirements or attributes follows. These customer attributes may consist of the weight of the paper, optical brightness, environmental friendliness, performance, dimensional stiffness, and cost. However, in considering the environmental friendliness of the product, the focus should be on the product life cycle assessment. Alternative products or substitutes are considered to determine which will create less environmental burden. The use of bleaching agents to clean recycled paper, conservation of forestry by using recycle paper, and other environmental components of the product are investigated. As shown in

Figure 8.3, any of these attributes may be present in any of the grades of paper selected by the customer. However, they may have different priorities depending on the grade. For example, when a consumer is interested in white waste paper as opposed to old newspapers, it is expected that attributes such as optical brightness and dimensional stiffness will be very important. Conversely, the cost will be higher for a white waste paper than for old newspaper. To improve optical brightness for example, the process of de-inking to remove contaminants will be more extensive in white waste paper. This will consume more time, energy and labor and may also require the addition of bleaching and other optical brightness agents to remove contaminants. These efforts may create additional environmental burdens that must be investigated and perhaps, possible substitutes evaluated. Similarly, when dimensional stiffness is very important, it may become necessary to combine a higher percentage of virgin pulp with the recycled fiber, thereby, making this paper less environmentally friendly. Taking a product life cycle assessment view, it is seen here that the use of white paper may create more environmental burden. For example, it may require high percentage of virgin pulp, use bleaching and de-inking chemicals that may potentially, become contaminants to the environment, create more waste disposal problem, and consume more energy resources such as water and fossil fuels. It is therefore, not sufficient to stop at this point. The environmental requirements for this product could be slated for further discussion. A separate framework such as the House of Quality may be created just for this criterion and alternatives or substitutes to the use of white paper looked at to determine the different options for white paper that could be integrated in the framework. Some of the options may include determining the minimum level of virgin fiber that should be present to achieve acceptable quality white paper. Thus, there are costs and benefits associated to whatever grade of paper the consumer needs. Once a particular group of paper type (i.e. Grade A) is selected and customer requirements are identified and prioritized, the next phase is to develop the design requirements. This involves designing for ease of manufacture. The manufacturer has to identify the design strategies that will enable it to satisfy customer requirements. Each of the attributes identified by the consumer can be broken down into a list of design requirements. For example, dimensional stiffness can be broken down to include the addition of virgin pulp, starch, stronger coating, and the use of less filler content, while optical brightness many include factors such as the use of an optical brightness agent and coating, longevity, bleaching, and de-inking processes. To produce environmentally friendly paper, we may look at issues such as the percentage of total recycled paper, percentage of post-consumer fiber, and the levels of chlorine-free or tree-free fiber present.

Once the key design requirements are identified by matching them with customer requirements, an optimal design plan is identified and the specific level for each design requirement is found. Sensitivity analysis also may be conducted on this design plan. The final phase is to evaluate the cost that may be involved and this should include both the direct and indirect costs. The direct cost should include known costs such as those that constitute the production cost, while the indirect cost should include environmental and social cost. It is necessary in evaluating the environmental and social cost to adopt a more systemic view of cost. This will involve the use of life-cycle assessment models. For example, the cost of the product should be traced through its life by tracing its pollution discharges and resources utilization as the product changes hands from producers to consumers. Since in recycling, we extend the life of the product by extending its usage and application, it is important that an economic input-output analysis becomes part of the life-cycle assessment. This analysis will consider the impact of the product in the different industrial sectors it passes through over its life cycle and by so doing; we can capture all the costs that may be related to the product. Of course, product life cycle assessment is often difficult to accomplish but it is a necessity in reducing a product's environmental burden.

Sometimes more than one grade of paper may be able to accomplish the desired needs of the customer. In such instances the final choice will be based on the grade that has the minimum total cost. Note again, that cost here includes the cost of environmental burden.

We shall illustrate this framework by the application of known models at each stage of the framework. The steps for conducting the green design are given in Figure 8.4.

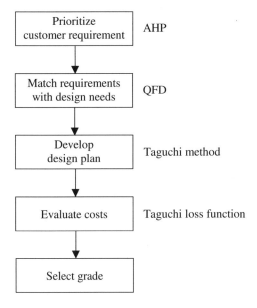

*Figure 8.4: Steps for Green Design*

## Design for Six Sigma

In order for today's companies to respond proactively to the needs of customers, it is imperative that processes are aligned to customer requirements and that such requirements are timely satisfied. Companies compete on several fronts but product quality is assumed and must be measurable. Six sigma is a rigorous approach to achieving quality and relies on effective data management and statistical application. The goal of six sigma is to apply statistics to data to identify sources of process or product defects, reduce variability, and achieve near perfection by aiming for zero defect. Conceptually, six sigma implies about 3.4 defects per million parts produced. Thus, quality is measurable and zero defects can potentially be achieved since this low number is proportionally zero. Six sigma offers a systemic and a systematic approach to identifying product or process problems by helping organizations to study their processes and workflows and enabling them to understand areas that need improvement. Effective and intelligent decisions can be made through the application of six sigma concept. In order to effectively analyze business processes, six sigma uses a five stage approach that is referred to as DMAIC. DMAIC stands for **D**efine, **M**easure, **A**nalyze, **I**mprove, and **C**ontrol. These stages are briefly discussed:

**D**efine – It is important for organizations to be clear on the opportunities they are faced with. Foremost in the six sigma approach is to define the aim or goal of the improvement activity. What is it that the organization really wants to improve on? A strategic objective may be to increase market share, reduce operating cost, achieve higher rate of return.

**M**easure – Measurement is essential in knowing when an objective has been achieved. Therefore, there is a need to establish reliable metrics or measurement scales to ensure that the defined objectives are being monitored. How does the organization know when higher rate or return has been achieved or when market share has been substantially increased? There needs to be a yardstick for measurement.

**A**nalyze – From measurement, we know the current status of the process or system. We can identify the gap between the current system state and the goal, and the objective is to minimize this gap so that the goal can be achieved.

**I**mprove – The process of improvement assumes that one understands the problems and the need to improve. Improvement requires finding better and more effective ways to do things and requires monitoring the new ways to ensure that indeed improvement was achieved.

**C**ontrol – Control provides a systematic way of institutionalizing an improved system to ensure that what works is put in place. Effective standards may be applied as a way of monitoring the system's performance so that it could be effectively tracked and deviations from established standards detected on a timely basis. The control however, is not close-looped but open to information and feedback that can be used to continuously improve and adapt the system. This way, the goal of zero defects can be effectively achieved.

The six sigma concept is applicable to environmental management systems (EMS). EMS pays attention to performance metrics. There are specific environmental goals, standards and targets that must be met. During product design, those environmental goals are specified and are tested to ensure compliance. However, a company becomes competitive when it attains six sigma in its performance metrics and becomes a world-class company to be benchmarked by others and its competitors.

Six sigma relies heavily on the application of statistical techniques and management. Therefore, it should be noted that a lot of statistical tools are used in the five stages of six sigma. Management is also crucial in the effective application of six sigma. Management of people, product, and processes are essential. When six sigma is effectively implemented, organizations are able to consistently achieve their environmental targets, improve their performance, reduce product and process variation, and satisfy the needs and wants of customers. The ultimate result of all these will be an increase in profitability, lower defects and higher quality, greater employee morale, and increased competitiveness. When these are achieved, environmental performance and productivity are improved.

## Measurement Issues in Six Sigma

The aim of six sigma is to achieve process improvement. Processes that are not compliant and are producing high levels of wastes, defects and pollution would be detected and corrective actions taken. Six sigma focuses on goals and results and enables the firm to look closely on how its products and processes can be aligned to meets it's environmental and quality goals. As we already mentioned, this is done by using a combination of statistical and management tools. By improving the sigma level, stakeholder requirements can be achieved. The use of DMAIC that is outlined above is to enable a systematic approach to focus on measures that are "critical to quality." These measures are often referred to as CTQ for short. CTQ does not necessarily have to focus on the intrinsic values of the product but should also consider environmental goals and targets that must be met for the product to be considered as meeting the quality goals. Six sigma can be applied to both physical and non-physical items as may be apparent in the ever-growing service industry. While for physical items, process performance is measured in defective parts per million (ppm), for non-physical items, it is measured in terms of defects per million opportunities (dpmo). Both defective parts per million and defects per million opportunities view environmental compliance as important in satisfying customer or stakeholders' needs. When there is less variation in the process or the process becomes more stable and consistent, smaller deviations are achieved. Smaller deviations imply higher sigma levels. Figure 8.5 presents a conversion between dpmo and sigma level.

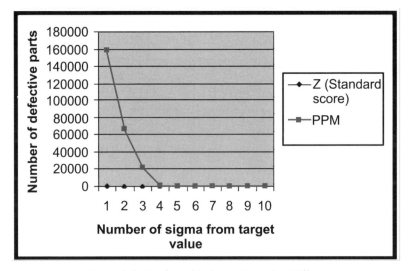

*Figure 8.5: Number of Defective Parts Per Million*

The customer focus adopted by six sigma shows that it actually works hand in hand with quality function deployment. While quality function deployment aims to design products and services that satisfy customer requirements as embodied in both product quality and environmental requirements, six sigma aims to achieve perfection in those products and services by reducing the variation that may exist through process improvement. In fact, the statistical and management techniques that are used to achieve six sigma include quality function deployment, failure mode and effect analysis, design of experiments, robust design, mistake proofing such as Poke Yoke and statistical process control [Goh and Xie, 2004]. Six sigma is an integral part of the product and service design stages. As products are designed for the environment through the product conception stage to the product's lifecycle, six sigma plays a key role. These techniques that are listed, are considered at the designing and monitoring stages of the product or services. They are not end of the line application techniques. As a result, Goh and Xie [2004] note that the statistical thinking involved in six sigma helps to make decisions that are based on facts. Six sigma relies on information gathering, and application of proven techniques to achieve process improvement. Finally, Goh and Xie [2004] characterized six sigma using "5W+1H." Specifically, this definition outlines the process of six sigma and how it leads to process improvement. They are adopted from Goh and Xie [2004], and Kuei and Madu [2003] and presented as follows:

WHY six sigma? Satisfaction of customers.
WHO does it? Structured top-down hierarchy or trained personnel.
WHAT is it? Statistical thinking using data to combat process variation.
WHERE is it? Standardized framework of "DMAIC."
HOW is it done? Software packages for information analysis.
WHEN is it done? Sustained effort via projects.

## Approaches for Six Sigma Projects

There are basically two approaches to implementing six sigma. We have already discussed DMAIC, which aims to achieve continuous improvement through reduction in variability of products or processes. One argument against this approach is that it may not react readily to changes. Thus continuous improvement on a process that is out of date will not enable the organization to compete effectively. Sometimes, breakthrough thinking is needed. A popular example is the effort to improve the Polaroid instant camera since digital photography has made the instant camera technology obsolete. The same applies in trying to improve a product that has passed its product lifecycle. Ultimately, it is important to understand what the customer wants as we have said many times in this book and then try to design six sigma quality into the product or process. So, the application of DMAIC will be satisfactory provided that the product or process remains capable of satisfying customers' needs. Its application is not innovative but aims at eliminating or avoiding defects from an existing product or process. Once the product or process is out of favor, applying DMAIC will not help the organization to become competitive. A new strategy must be adopted. That strategy is the second approach, which is referred to as DMADV (design-measure-analyze-design-verify) or IDOV (identify-design-optimize-verify). DMAIC is based on continuous improvement, while DMADV or IDOV is based on reengineering of the process. This radical approach is also often referred to as design for six sigma (DFSS). Harry and Schroeder [2000] claim that once organizations have reached five sigma quality levels (i.e. 233 defects per million opportunities), they need to design for six sigma to surpass the five sigma. As noted by Banuelas and Antony [2004], there are not enough data to substantiate the claim even though some authors have supported it. We adapt from Nave [2002] and Harry and Shroeder [2000] the assumptions made by DMAIC and IDOV as presented in Table 8.1.

*Table 8.1: Contrast of DMAIC and IDOV or DMADV*

| Dimension | DMAIC | IDOV or DMADV |
|---|---|---|
| Objective | Adopts current product or process design as correct and economical and aims to minimize the variation. | Aims to achieve resource efficiency through design to satisfy current customer and market requirements. |
| Process capability | Current product or process design is capable of satisfying customer needs. | Aims to achieve more than five sigma quality by producing higher yields regardless of complexity and volume. |
| Design | Customer needs and market requirements are satisfied by the current product configuration. | Focuses on "robustness" and capable of withstanding adverse conditions. |
| Flexibility | Customer needs are currently met with current configuration and design. | Has a high focus on customer demands and adapts to such demands. |

Thus, the main distinction between these two approaches is the fact that DMAIC accepts the existing state of the system and works on improving it through reduction in product or process variation. Conversely, IDOV or DMADV questions the existing process and tries to develop a more proactive look. It is based on designing six sigma quality into the product or process, in order to continuously satisfy customers' needs. It has a focus on challenges and innovation to meet the dynamic changes in the market place.

In fact, the IDOV or DMADV approach is similar to the argument proposed in Goh and Xie [4] who countered that rather than DMAIC focusing on error or defect avoidance, that six sigma can be extended to include a systems perspective and strategic analysis. Systems perspective will focus on identifying appropriate system boundaries and performance indices. This will go beyond focusing on the product or process variation but to understand the macroeconomic environment that influence the "critical-to-quality" measures. Such macro view considers other factors that could potentially influence product quality and environmental issues are important in such analysis. The CTQs have to be frequently reviewed to recognize when new measures emerge and when to adapt the list. The strategic analysis on the other hand is scenario-driven to address "what if" questions that may arise in the dynamic market place. Product lifecycles are short these days. Technologies easily become obsolete, and customer needs and wants are fast evolving. It is important that organizations understand this dynamism and be able to react appropriately to respond to the needs of the market place.

### Six Sigma and QFD

As we have already defined, QFD is used to translate customer requirements into a set of prioritized targets that can be focused on when improving products, processes and services to satisfy the customer. These customer requirements are referred to as critical-to-quality measures in the six sigma literature. Six sigma is a highly structured approach that has a strong focus on the bottom line. Achieving almost a zero defect in defects per million parts or defects per million opportunities should translate to financial gains for the organization and make the organization competitive. So, both QFD and six sigma aim to make the organization more competitive and successful by aligning the organization's strategy to focus on its customer and by developing means of measuring process performance improvements. One of the approaches for six sigma we discussed is the design for six sigma (DFSS). DFSS is based on listening to the voice of the customer to identify and prioritize the critical-to-quality measures that are important to the customer. These CTQs must be integrated into product and process design if the needs and wants of the customer are to be satisfied. The CTQs will be identified and prioritized by the members of the cross functional teams. This way, the organization focuses on the key CTQs and the methods of six sigma can be applied to achieve both robust design and minimize the variations in those key components. Indeed, with the application of QFD, six sigma teams can optimize their resources by targeting the CTQs that are high on importance rating to the customer. An improved process will have no value if it does not focus on what the customer wants. The customer knows what is critical to achieve his or her satisfaction. These CTQs can only be understood by listening to the voice of the customer. Thus one can emphatically state that before six sigma could be applied, CTQs must be first identified and prioritized. The priorities attached to these CTQs would also influence the specification limits that should be set. The most important CTQs would definitely be set to higher specification limits such as six sigma because it would be essential to significantly reduce variation in such CTQs and ensure their consistency and stability. Another aspect of the QFD is the correlation matrix, which appears as a roof in the House of Quality. By identifying potential conflicts between design requirements, emphasis is placed on the most important design requirements. Again, these become the design requirements where serious efforts must be made to achieve six sigma levels. So, rather than the six sigma team using its resources to fight fires all over, it can from QFD identify both the most important customer and design requirements and focus its efforts on those. This way, a product that meets customers' needs can be designed and produced on a timely fashion. Such products will also help the organization compete effectively and become profitable.

## Implementing Six Sigma with QFD

• The goals of six sigma and QFD are not quite different even though different approaches may be followed. Ultimately, the aim is to satisfy customer needs so that the organization becomes competitive and profitable. To realize this goal of achieving customer satisfaction, top management participation and support are needed. Top management plays a critical role of creating and cultivating an environment that will support innovation and creativity that are essential for the implementation of QFD and six sigma concepts. Such an atmosphere will challenge existing hierarchical structures in the organization, division of power and authority and breakdown the barriers that may hinder innovation and change. Top management also bears the full responsibility of allocating needed resources to ensure that the implementation of six sigma within QFD is successful. One area that demands resource allocation is training and the use of the best brains in the organization to effectively adapt to six sigma. The training requirement has often hindered the implementation of six sigma in smaller companies. For example, big companies like Motorola spent $170 million dollars between 1983 and 1987 on worker education that focused on quality issues such as quality improvement and designing for manufacturability. Yet, the training program has to be properly designed for it to be successful. One of the early problems encountered by Motorola was that it followed a bottom up approach in its training by training lower level employees on statistical process control without providing them remedial education. When the training flopped, it was difficult to turn to top management who had not been trained to provide help. Consequently, Motorola established Motorola University to provide training to its executives. It is therefore imperative that everyone be trained and that top management takes the lead in training. However, training does not have to be a drag on corporate resources. It is important to benchmark organizations that have efficient training programs. General Electric appears to have successful training programs that is well structured and takes lesser time. For example, while it takes GE 16 to 20 weeks to train a black belt (a basic level of six sigma certification), it takes Motorola a minimum of one year.

- It is important to listen to the Voice of the Customer. There must be a free flow of information where customer needs and wants are identified. The critical to quality issues are known and the design strategies are developed to address such issues. Resources are efficiently utilized when it is clear which customer requirements are more important and which design requirements need tighter specification as the six sigma. Thus implementation will be more effective when the focus is on the critical design issues.

- Training needs at the different levels of the organization must be assessed. But before the training, employees must understand the new mission of the organization to achieve customer satisfaction and how training on both QFD and six sigma could help realize such goals. It is important that remedial skills are provided at all levels of training to achieve both literacy and statistical understanding and must involve a top-to-bottom approach.

- As we have mentioned above, there are two approaches to six sigma namely DMAIC and DMADV (or IDOV). Process implementation must involve a two-prong monitoring system. One will focus on DMAIC where incremental and continuous improvement is achieved once the current processes and products continue to satisfy customer and market needs and the DMADV or IDOV will involve reengineering of the entire product or processes when the current system is incapable of meeting customer and market needs. This way, the organization is adopting both systemic and strategic perspectives and would be able to design for six sigma and adapt the design strategies as customer requirements change.

- Cross functional teams would have to be empowered and remain actively involved in evaluating and monitoring business processes, scanning the environment and developing strategies to continue to respond proactively to the dynamic environment.

## Lean Six Sigma

Many authors especially practitioners have tried to distinguish between Total Quality Management (TQM) and Six Sigma and have often confused the grey areas between the two. In fact, some distinctions have often misrepresented what TQM is, in favor of the new focus on Six Sigma. While Six Sigma is in vogue in many corporations such as Motorola, GE, Seagate Technology, GlaxoSmith Kline, Ratheon, and a host of others, critics argue that six sigma is another management fad that may ultimately not meet the expectations of the adopting companies.

Even successful companies that have tried to implement these concepts such as TQM, Business Process Re-engineering and Six Sigma have recorded failures [Easton and Jarrell, 1998]. There is also the problem of sustaining initial process improvement successes [Sherman et al, 1999]. Further, business performance has not always improved with successes in these programs as evident with the layoffs and lack of profitability by Motorola in 1998 [Basu 2004].

Many have even proposed Lean Six Sigma as an extension of Six Sigma. Lean Six Sigma is simply an integration of Six Sigma with Lean Production. A lean organization focuses on value management and aims to eliminate non value-added activities. The elimination of wastes helps the organization to become more productive and efficient, and able to achieve operational excellence. Lean Six Sigma application covers the entire value delivery chain which extends to vendors and suppliers and adopts customer focus. It helps organizations to become agile and to respond rapidly to their dynamic environments. Organizations adopting lean six sigma often focus on sustainable practices and aim to reduce environmental burden.

## An Example of Six Sigma Application

In technical terms, sigma ($\sigma$) denotes variability or standard deviation of an attribute or customer requirement of interest. In statistical process control, specification limits are provided and a critical assumption is made that the process is stable and behaving according to the normal distribution. When a process is stable, its behavior can be predicted. With the QFD phase, the customer requirements are identified and used as a basis in developing quality characteristics to measure. It is important that the process is capable of satisfying these quality characteristics in order to satisfy the needs and wants of the customer. Six sigma helps to ensure that this is done. As we have already stated, this approach will ensure that only 3.4 defects are observed per million parts or opportunities.

There is often a misunderstanding on what the six sigma process means. A major assumption in the way six sigma concept was developed is that the shift in the process mean is within 1.5 standard deviation and that the process mean is 4.5 standard deviations from the closest specification limit and 7.5 standard deviations from the farthest specification limit. In fact, the process mean is 4.5 standard deviations from it closest specification limit and not 6 standard deviations as the term 6 sigma would tend to imply. It would probably have been better and easier to refer to this approach as "four and a half sigma" rather than "six sigma." Mitra [2004] provides a good description of the statistical foundation of six sigma and notes that the expectation that a process stays within 1.5 $\sigma$ of the process may vary from process to process. In other words, each organization has to understand its process spread and mean and apply them in finding process shift.

*Table 8.2: Computation of Defects for One-Sided Specification Limit*

| Z (Standard score) | Normal Probability | Complement | PPM | PPB |
|---|---|---|---|---|
| 1 | 0.841344746 | 0.158655254 | 158655.3 | 158655253.9 |
| 1.5 | 0.933192799 | 0.066807201 | 66807.2 | 66807201.27 |
| 2 | 0.977249868 | 0.022750132 | 22750.1 | 22750131.95 |
| **3** | **0.998650102** | **0.001349898** | **1349.9** | **1349898.032** |
| 4 | 0.999968329 | 3.16712E-05 | 31.7 | 31671.24184 |
| **4.5** | **0.999996602** | **3.39767E-06** | **3.4** | **3397.673134** |
| 5 | 0.999999713 | 2.86652E-07 | 0.3 | 286.6516541 |
| **6** | **0.999999999** | **9.86588E-10** | **0.0** | **0.9865877** |
| 7 | 1 | 1.27987E-12 | 0.0 | 0.001279865 |
| **7.5** | **1** | **3.18634E-14** | **0.0** | **3.18634E-05** |

Table 8.2 shows the basic concept of six sigma. If the six-sigma process is centered at the mean, then the process would produce one defective part per billion (PPB) for each specification limit or two defective PPB for both specification limits. Similarly, if we apply the concept that six sigma is 4.5 from its closest specification limit and 7.5 standard deviation from its farthest specification limit, we see that 3.4 defective parts per million (PPM) are produced. We see from Figure 8.6 when we contrast the standard 3 σ to 6 σ, that six sigma would tend to produce more consistent products than three sigma. Thus, the higher the number of sigma, the more capable and consistent the process becomes. A simple example would be a mail delivery company that checks its on-time delivery using three and six sigma respectively. A three sigma operation would expect 1350 late deliveries per million mails and for a truly six sigma operation, the organization would expect 2 late deliveries per billion mails.

Thus, QFD and Six Sigma are important tools in designing for the environment. These tools will help to understand the stakeholders' needs for the environment and to develop achievable standards on how environmental quality can be achieved. Several national and world agencies on environmental management have established standards on emissions, pollution and waste disposals for corporations and governments alike to follow. The use of lean six sigma and six sigma will help in establishing targets and standards to detect when compliance is being met. It is very important that results are produced and are also monitored and evaluated. More importantly, design for the environment should start at an early stage in the product development process. This is the more reason why tools like QFD have become increasingly important. It is necessary that all stakeholders participate in evaluating their needs and requirements and matching those needs and requirements to design strategies.

In today's complex environment, there are many competing needs that must be designed in the product. The concern for the environment has increasingly taken a bigger role and value cannot be created without paying attention on how the product or service interacts with its environment. The need to lessen environmental burden should be a top priority in any product or service design hence the need to design for the environment.

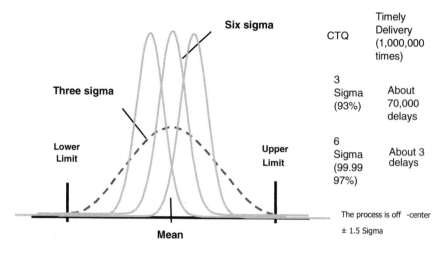

*Figure 8.6: Three Sigma vs. Six Sigma*

## References

Banuelas, R., and Antony, J., (2004), "Six sigma or design for six sigma?," *The TQM Magazine*, 16 (4): 250-263.

Basu, R., "Six Sigma to FIT sigma: The New Wave of Operational Excellence," http://www.onesixsigma.com/_lit/white_paper/fitsigma_RonBasu.pdf, downloaded on September 27, 2004.

Easton, G., and Jarrel, S., (1998), "The effects of Total Quality Management on corporate performance," *Journal of Business* 71, 253-307.

Goh, T.N., and Xie, M., (2004), "Improving on the Six Sigma paradigm," *The TQM Magazine*, 16 (4): 235-240.

Harry, M., and Shroeder, R., (2000), Six Sigma: The Breakthrough Strategy Revolutionalizing the World's Top Corporations, Doubleday, New York, NY.

Hoerl, R.W., (2001), "Six sigma black belts: what do they need to know?," *Journal of Quality Technology*, 33 (4): 391-406.

Kuei, C.H., and Madu, C.N., (2003), "Customer-centric six sigma quality and reliability management," *International Journal of Quality and Reliability Management*, 20 (8): 954-964.

Mitra, A., (2004), "Six sigma education: a critical role for academia," The TQM Magazine, 16(4): 293-302.

Nave, D., (2002), "How to compare six sigma, lean and the theory of constraints," *Quality Progress*, 35 (3): 73-79.

Sterman, J.D., Keating, E.K., Oliva, R., Repenning N.P., and Rockart, S., (1999), "Overcoming the improvement paradox," *European Management Journal*, 17: 120-134.

Bovea, M.D., and Wang, B., "Integration of customer, cost and environmental requirements in product design: An application of Green QFD," undated.

Madu, C.N., House of Quality (QFD) in a Minute, 2nd edition, Fairfield: CT., Chi Publishers, 2006.

Madu, C.N., Kuei, C-H., Madu, I.E., "A hierarchic metric approach for integration of green issues in manufacturing: A paper recycling application," *Journal of Environmental Management*, 2002.

## Chapter 9

# Manufacturing Strategies: Agile, Lean and Flow Manufacturing

In this chapter, we discuss some of the new manufacturing technologies and how they could help improve quality and productivity and reduce societal environmental burdens. We shall focus specifically on agile manufacturing, lean manufacturing, and flow manufacturing. These techniques also benefit from the application of just-in-time inventory system, total quality control, and total productive maintenance. These relationships are discussed in the chapter.

### Agile Manufacturing

Agile manufacturing is a new manufacturing philosophy that originated from the Japanese auto and consumer electronic industries. The aim is to inspire to achieve total flexibility with high quality at minimum costs. Agile manufacturing is intended to transform the traditional manufacturing practice to something more efficient and productive and responsive to the dynamic needs and demands of the customer. The basic thrust of agile manufacturing was identified in Keen's "Agile Manufacturing" [2001]. We shall expand the discussion to include the following:

- Reducing dependency on the economies of scale—unlike traditional assembly line system where volume drives production, agile manufacturing encourages flexibility and therefore, variations in the basic product design which counters the focus of the traditional flow shop operation. The mass production system creates waste. It assumes commonality in the needs of customers and items are made to stock. These items may not necessarily satisfy the needs of the customer. Unsold items are scrapped as waste thereby encouraging material and energy consumption. Furthermore, such a system of production can hide defects. This would mean that in the long run, the customer would be able to detect such defects and return the item for rework or replacement. Such rejects places heavy demands on natural resources and energy and material requirements, thus leading to creation of wastes. Further, mass production encourages stock piling of inventory. Inventoried items place demand on space, energy and materials. Agile manufacturing uses new adaptive manufacturing technologies such as ["Agile manufacturing," 2001]:

    a.   Rapid prototyping (RP)—these are a class of technology that are used to develop prototypes and may include three dimensional computer-aided designs. The use of rapid prototyping is a flexible approach that would enable investigation of alternative manufacturing process to address "what it" scenarios. Consideration of environmental impacts could be integrated in the design process to ensure that products that achieve environmental quality are designed and produced.

    b.   Rapid Tooling (RT)—this consists of the additive and subtractive processes. The additive processes use advanced methods of making tools based on the RP technology while the subtractive processes use the advanced methods of making tools that are based on the milling technology. Both methods rely on the use of digital database. The rapid tooling also ensures that the right processes are used. The right processes should meet environmental compliance and be able to meet the environmental guidelines of the manufacturer.

    c.   Reverse Engineering (RE)—this is a hybrid of approaches to reproducing physical objects. It may involve manual drawings, documentations, computer-aided approaches or a mixture of both. Reverse engineering supports the use of any method or approach that may be necessary to reproduce an item. Reverse engineering is adaptable to different methodologies. It could therefore, be adopted to design for the environment.

- Low volumes are produced at competitive prices—this seems to encourage small orders rather than mass production operation. The small volume enhances environmental quality checks.

- Decentralized mini-assembly plants—the notion of maintaining a centralized assembly line system that produces basically identical products is replaced with a decentralized mini-assembly plants that are situated near demand centers. This helps to cater to the needs of the different demand centers but more importantly, satisfy the goal of just-in-time delivery. It therefore, helps to reduce inventory levels and the costs associated with inventory thus adding more value to the customer. With JIT, it is easier to track problems when the product fails to meet either product or environmental standards. The root of the problem can easily be analyzed. Operator's pay serious attention to their work since JIT would not hide defects. Environmental compliance could therefore, be achieved.

- Flexibility—Flexibility is important in agile manufacturing. Different configurations of the product are encouraged and offered. With flexible system, it is easier to continuously improve by adapting the product or process to meet new standards.

- Worker motivation—Worker motivation is enhanced because the worker is no longer bound to routine tasks as defined in the traditional assembly line system. Rather, the worker provides input in configuring his or her work operation and is not bored by repetitive tasks. The fact that the worker now enjoys work helps to reduce human error which may be a major cause of environmental problems. Further with rapid changes in product designs and the flexibility of the process, workers can have significant input in designing products that meet environmental standards.

- Product design—the product design is based on listening to the voice of the customer. Rather than designing a product or service as perceived by the engineer, the customer is an active participant in designing the product or service to meet his or her needs. Thus, customer satisfaction is high. The product is also designed for the environment and follows guidelines to achieve environmental standards.

- Supply chain management—supply chain management is emphasized, as there is need for suppliers, vendors, and distributors to share the same goals as the manufacturer and meet both its quality and environmental standards. Thus, a transformation of the traditional type of relationship with suppliers is needed. Supply chains add value to the product by recognizing the need to achieve the environmental goals of the stakeholders.
- Coordination—to achieve the goals of agile manufacturing in terms of quality, speed of delivery, and flexibility, good coordination and management of information is required.

Agility is a rapid response to the changing operating environment of the firm. Businesses operate in an atmosphere of intense competition that is subject to challenges and uncertainties. The business environment of tomorrow is unknown let alone the future. The firm is increasingly facing many threats emanating from the following:

- Customer needs and expectations are very dynamic and evolving.
- Intense competition from new entrants into the industry and even from firms in other industrial sectors.
- Globalization of businesses while opening up new markets, has led to new competition with different rules for the game. The new competitors often come with different strategies and philosophies that are difficult to predict.
- The rapid proliferation of new technologies and new products has increased Research and Development costs and led to drastic shortening of the product life cycle. Market niches are no longer guaranteed, safeguarded or protected from competition.
- Increased number of stakeholders and interest groups with their own agendas i.e., environmental interest groups, diversity groups present new challenges that businesses must respond to.

With all of these elements, the need to achieve environmental quality is salient. New laws and regulations on emissions and management of environmental burdens have influenced the development of new products and processes. Companies that have adapted to this new trend have achieved competitiveness. With new technologies and new product designs, wastes and environmental emissions and pollution are minimized. In fact, controlling of waste is a competitive weapon that creates strategic advantage to firms.

In order to meet up with the demands of agile manufacturing, the firm must be flexible, robust, and ready to adapt. It must readily scan its environment to identify new opportunities and challenges and align its strengths to meet these rapid changes in its environment. An agile company must be in continuous movement with time to respond proactively to changes in manufacturing practices, technology, natural environment, and customer requirements. The key is to rapidly adapt to unexpected and unpredictable changes in the organization's operating environment. This is not easily achieved since rapid adaptation may often conflict with long-term organizational strategic plans and may also not align with organizational culture and climate thus disrupting the organization. However, this is the only way that the organization can remain competitive, and be able to satisfy the needs of its customers. There are five key areas that would specifically influence how the organization responds to its operating environment. These are the organization itself, management of change, product development, the natural environment, and relationship with customers and suppliers. We shall briefly discuss each of these.

## The Organization as a Change System

To respond rapidly to agile manufacturing, the organization needs to adapt and change. The traditional hierarchical structure of the organization would have to be transformed into a more fluid system with openness for sharing ideas, information, and knowledge. This would require a transformation of the organizational structure to recognize the mutual dependence between the different functional and business units of the firm. More so, the firm must appreciate its operating environment and recognize the importance of the information flow and two-way relationship it has with its operating environment. This would enable it to respond rapidly to the needs of its environment.

The organizational structure must be flexible to support acquisition and use of new information to update existing knowledge. Power and authority would have to be delegated to more people so decisions can be made on a real time basis. In other words, strategizing of decisions should be spread across the firm and not rest solely with top management. Decisions regarding the different processes that interact with the firm have to be distributed and coordinated so that the firm can respond rapidly to its dynamic environment. Such decisions may include process selections, suppliers/vendor selections, partnerships with competitors and suppliers, and customer relationships. Furthermore, a more coordinated and integrated decision-making process that includes more active participants or stakeholders should be encouraged while employees are empowered to make work-related decisions.

Such coordinated efforts would help in identifying members of the supply chain that would meet the firm's environmental goals. Furthermore, sharing of information on a timely-basis would help to effectively select the right process and to design for the environment. Critical environmental information could be lost, misinterpreted or not timely applied in a hierarchical structure. Furthermore, since lack of compliance could spell problems for the firm, employees may be uncomfortable exchanging information on environmental compliance when they perceive lack of support or authority in the matter. The lateral structure shown above supports cross-functional teams. Employees as members of this team feel empowered and are able to contribute. They see participation in such a team as an expected job function to enhance their work and deliver products or services with high values.

## Management of Change

To deal with agile manufacturing, a firm must develop its core competence in management of change. Change is inevitable and unpredictable and may often be clouded with uncertainties. Change is also a major threat to the organization's survival and yet may provide the best opportunity for growth and competitiveness. Organizational change is complex because it deals not only with the organizational structure and processes, but also the human element. It is not always easy to get people to change attitudes, values and culture that are acquired over several years. In fact, the ability to manage change will depend on how able the human processes will adapt to change. This may require a great deal of training and counseling to get employees to accept change as a necessity and vital to the survival of the organization. Agile manufacturing seeks a re-engineering of the traditional manufacturing process and such radical change cannot take place without a challenge from some traditionalists. It is therefore important to develop a proactive strategy to get everyone on board when agile manufacturing is introduced. Information sharing will be important and participatory decision-making will expose employees to the dangers of not adapting to the new process. A rapid response to change and developing strategic initiatives to manage change in the entire organization will enable the organization to respond proactively to its dynamic environment.

## Product Development

Changes are needed in product development. The ultimate aim of the product is to satisfy customer needs. Therefore, the customer should be a participant in deciding how his or her needs could be satisfied through product design and development. In addition, there are several stakeholders who are influenced by the product or influence the successful introduction and marketing of the product. These stakeholders are active participants and should take part in the product design. Product development should pay attention to a host of factors including quality, price, value, and speed of delivery, mass customization, environmental content, and product stewardship. Clearly, these factors could affect the successful introduction of the product. The product must be designed not only to meet product quality requirements but also, to satisfy environmental standards. It is important also to identify what competitors are doing and conduct a SWOT (strengths, weaknesses, opportunities, and threats) analysis to identify ways to continuously improve the product and beat competition.

## Natural Environment

A major strategic issue facing organizations today is how to optimize the limited natural resources by limiting the use of nonrenewable resources and minimizing waste. The lean manufacturing strategy could help in this regard by focusing on substitute components that have or create less environmental burden, taking a product stewardship, changing the pattern of consumption, and focusing on environmentally conscious manufacturing. The protection of the environment is a major challenge facing organizations today. Organizations need to develop socially responsible strategies to support the natural environment.

## Customer/Supplier/Vendor Relationship

The customer is the essence of any business organization. Without a customer base, the organization has no purpose and will not meet its mission. The firm must focus its strategies to listen and address the needs of the customer. Customer loyalty can only be earned when the customer is continuously satisfied. Rapid response to the changing environment is actually a response to the dynamic and the evolving needs of the customer. The firm must therefore, understand and appreciate what the customer perceives as important in satisfying his or her needs.

The supplier also plays a critical role in satisfying customer needs. The relationship between the firm and its suppliers have undergone significant transformation to building a mutual relationship that is based on trust and support for each other's activities. This mutual relationship has helped both parties to work toward a common goal to satisfy the needs and wants of the customer and should be fostered.

Customer/supplier/vendor relationship is helpful in product management and subsequently in environmental protection. The interaction and cooperation between the different intermediaries in product cycle management have helped to significantly reduce the amount of environmental pollution. If we trace back the Kodak SUC success story that was presented in Chapter 2, we see that it is a continuous loop that works because each of the value channels is dependent on the other. The customer or end user depends on the photofinishers to help with the processing of the camera; the manufacturer of the camera depends on the photofinishers to collect the camera from customers for processing, etc. There is need for communication and exchange of information between all the parties for this to work effectively. Figure 9.1, shows the cyclic loop of this relationship.

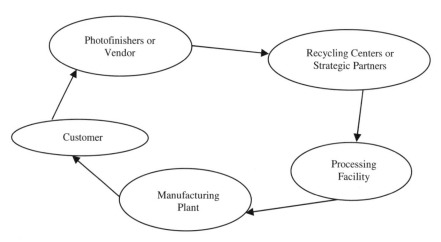

*Figure 9.1: Value Chain Loop*

## Agile Manufacturing vs. World Class Manufacturing

There are many other manufacturing strategies that have proliferated since the 1990s that support environmental management. Notable among these are mass customization and lean manufacturing. There is a tendency to log these other strategies with agile manufacturing as if they are all the same when in fact; they differ remarkably ["Agile

Enterprise/Next Generation Manufacturing Enterprise," Maskel, 2001]. Agile manufacturing can only be defined in the context of uncertainty and unpredictable changes thereby distinguishing it from the other manufacturing strategies. Mass customization on the other hand, while dissuading the notion of mass production and allowing for low volumes with different design changes to be fulfilled, it still relies on modular and standardized components to meet customer needs. Although it allows for a wide variety of product changes, the customer is often restricted to the list of options that are available. Outside this options list, mass customization may not respond. Dell Computers is a classical example of a company that uses mass customization. Dell Computers on its web site offers the customer the opportunity to build his own computer starting from a basic standard defined by Dell. Thus, the options open to a customer is confined within the boundaries and limits that are already established and the customer's add-ons can be applied in a modular setup. Certain changes that may be possible through agile manufacturing that would be application-specific are not easily made in a mass customization system. However, with mass customization, optimal use of materials is made since parts could be interchangeably used for different models. Waste is contained since the product is built at the time of order so there is less obsolescence of parts or products. Inventory and its associated environmental costs are controlled since it is the customer's demand that drives the order. Environmental wastes and pollutions are substantially controlled by focusing on meeting exactly the needs of the customer. Agility specifically, is the ability to accommodate and manage change. It thrives on the challenges and opportunities that the dynamic environment would present.

Conversely, lean manufacturing deals with an environment with little or no uncertainty. The environment is relatively stable and predictable and the manufacturer is in full control. In fact, a major objective of lean manufacturing is to limit uncertainties and any potential variations that may exist in the production process. Lean manufacturing therefore, cannot effectively handle future events that often deal with lack of information and greater degree of uncertainty and unpredictability. While lean manufacturing may be amenable to some form of standardization, agile manufacturing is dynamic and must respond likewise to its unpredictable and uncertain environment. These uncertainties make the environment for agile manufacturing more complex and more challenging to deal with. An agile manufacturer is a futurist that has no boundary but responds rapidly to the needs of its environment. As demands change so will the manufacturer change to meet the new demands. Presently, products become obsolete rapidly due to the rapid proliferation of new technologies and new products, and shorter product life cycles. It would be myopic to perceive that any product can last in the market for a significant duration without major changes in its design and intended use. Thus, an agile manufacturer must respond by introducing new products and even entering into other business areas that may present new challenges and opportunities. For example, personal computer manufacturers facing falling demand in the PC market are shifting into the manufacture of computer storage devices, computer game consoles, and even application software programs.

Agile manufacturing can be perceived as in constant flux of change. It requires re-engineering of the entire organization and its processes while the other manufacturing strategies thrive on stability and would do better with continuous improvement. The ability to adapt and manage change is the key to achieving agility and such requires radical changes in an organization's structure, culture, processes, strategy, and philosophy.

## Lean Manufacturing

Lean production is a manufacturing strategy by the Toyota Production System that is based on adding value through the elimination of waste and incidental work. This strategy strives to achieve the shortest cycle time, high quality, low cost, and continuous quality improvement. It is based on the fact that the business environment is unpredictable and very dynamic. Therefore, it is important to maintain a stable production system that will provide high value at minimum cost and waste. The objective of lean production is to streamline the manufacturing process by evaluating the entire process from product design stage to the product delivery and consumption stages to identify the value added at each stage and the wastes that can be cut out at each stage. The goal is to minimize waste that may result from inventory, material, inefficiency, and quality at every stage of the production process. Lean manufacturing helps to shorten the product cycle time and to design and deliver products that are flexible and able to satisfy customer needs at the lowest possible cost with high quality and as quickly as the customer demands it. Supply chain management also plays a major role in achieving the goal of lean manufacturing since a coordinated network of suppliers is necessary to achieve the just-in-time inventory system that is maintained.

A synchronized production system is maintained to respond rapidly to customer demands. This system effectively manages work and equipment utilization and scheduling utilizing real-time information. This rapid response to customer demands, increases productivity, equipment utilization, reduces cycle times, and minimizes waste due to reduced number of scraps, rejects, and reworks. An integrative approach to information sharing between the stakeholders ensures that the system responds rapidly to customer needs and that all the business units are coordinated to achieve customer satisfaction.

Baudin [1997] identified five guiding principles for lean manufacturing. These principles could help articulate the importance of lean manufacturing in a business enterprise. We shall expand the discussion on these principles below Baudin [1997]:

- People are the main drivers of productivity—the main premise of this principle is to invest on people as the primary asset of an organization. With increase in automation in our world today, there is a tendency to perceive machinery as smart, intelligent, and productive. However, machineries are still easier to replace while people are indispensable. The knowledge and information embedded in people could be a major source of competitiveness for an organization. According to Baudin, people offer "both muscles and brains." A firm can achieve competitive advantage by taking advantage of what people have to offer. To achieve this, a work environment that fosters trust and respect, and challenges the worker has to be created. Employee satisfaction is the key. Employees have to be viewed as critical members of the business enterprise and not perceived as disposable commodities that can be easily replaced. The workers need to be motivated, allowed to achieve their self-actualization through work, and empowered to make critical work-related decisions. Their sense of belonging and association with the firm has to be enhanced and they have to perceive their self-development and growth through the organization. When employees identify with the organization, they develop pride and joy from working for the organization and are able to improve their productivity and quality. Issues relating to environmental management may often be perceptual. Employees live in the natural environment and interact and socialize with others. Certain information and experiences gained from this interaction may help in designing an environmentally conscious process or work environment.

- The key to profits is on the shop floor—adequate attention needs to be paid to how the shop floor is run. Many of the wastes that are incurred in lean manufacturing can be traced to the shop floor. Wastes may be in the form of material, energy, or manpower resulting from poor quality, pollution, or redundant activities. It is therefore, necessary that efforts are made to understand the tasks and activities at each stage of the manufacturing process and balance their value contributions to their waste creation. Significant wastes could be trimmed by redesigning how the workplace is laid out and how activities and tasks are carried out. Furthermore, an efficient design of the shop floor may help to improve quality and productivity, by cutting down on scraps, rejects, and reworks, and help the organization to focus on doing things right the first time.

- All manufacturing is repetitive—this concept is based on the premise that most manufacturing operations at least for a product or within a product family are repetitive. There are only few customized operations. Even in a mass customization system, there is a significant degree of repetitiveness. Organizations adopt flexible manufacturing systems to respond to varying customer demands that may deviate from the basic product design. Small-scale job shops may be maintained to satisfy these specialized needs. Flexibility allows the manufacturer to take advantage of these small orders without incurring the added cost. Waste is also minimized as it is harder to hide defects in smaller jobs.

- The work must flow through the shop—given the varying needs of customers and the need to customize both products and services, work must be arranged based on a job shop flow rather than on a flow shop system. In other words, people and equipment should be arranged according to the tasks they perform and work orders should be scheduled to pass through the job shops. This will help to deal with bottleneck problems encountered with the flow shop system, and will make it economically feasible to support small job orders. A "pull-driven" rather than a "push-driven" system for the transfer of parts and materials along the plant is supported. The pull system encourages the production of products according to actual orders or demand rather than based on market forecasts. This is the JIT system which supports environmental management.

- Improve, don't optimize—Management must begin to de-emphasize optimization and focus on achieving continuous improvement. There is no "optimum" or "best" solution when dealing in a dynamic and unpredictable environment that is faced with a lot of uncertainties and constraints. Every effort should be made to continuously improve the process, work processes, employees, organizational processes, productivity, and quality. There is no end in achieving continuous improvement.

Lean manufacturing can be viewed as antithetic to the traditional assembly line operation that relies heavily on the ability to forecast independent demand and the use of mass production system. Lean manufacturing is based on make-to-order and the use of a pull-driven system. Rather than offering repetitive and monotonous tasks to employees, it empowers them to make decisions about their work. There is a high focus on flexibility, quality, and productivity improvements. However, as we have already stated, lean manufacturing should not be confused with agile manufacturing. Lean manufacturing benefits from a well coordinated management information system where both the manufacturer and the supply chain network share information on a real time basis and are able to respond swiftly to changes in the environment.

## The Thrust of Lean Manufacturing

The thrust of lean manufacturing can be found in three popular manufacturing techniques namely the Just-In-Time (JIT) system, Total Quality Control (TQC), and Total Productivity Maintenance (TPM) ["Lean manufacturing,"2001]. We shall briefly discuss them.

## Just-In-Time

The essence of JIT is to reduce inventory to bare minimum by delivering parts and materials exactly when they are needed. JIT ensures that parts and materials are delivered from upstream activities to downstream activities so that inventories do not accumulate at any stage. Therefore, a consumer downstream would have to signal for order before it could be delivered. These signals are known as *Kanban* signals as coined by the Toyota Production System. A fixed buffer is maintained at each workstation. When the buffer is depleted, a kanban is sent to the producer who replenishes the buffer with a new order. This process synchronizes upstream and downstream activities to ensure smooth transfer of parts and materials at minimum inventory levels at workstations. The JIT system is known as a *pull system* or *pull manufacturing* since it is triggered by actual rather than forecast demand. The major problem in the use of a pull system can be traced to the effective coordination of all workstations and being able to correctly calculate buffer requirements at each workstation. This system increasingly becomes complex as the supply chain network increases. However, the availability of advanced software technologies and the application of electronic commerce are helping manufacturers to better coordinate supplier's inventory system and synchronize them with the manufacturer's internal needs. This would help achieve the JIT goal.

The use of JIT helps to focus on quality achievement. It is difficult to hide defects. With mass production system, defects could be hidden since items are stored for usage at a later time. JIT would therefore, support environmental compliance.

## Total Quality Control

Our modern view of quality management is shaped around total quality control. TQC is based on a company-wide strategy to achieve quality by minimizing scraps and rejects and by achieving customer satisfaction. Much has been said about quality management in the literature and how quality is directly associated to productivity improvements and cost control. The main idea behind quality practice today is that it is everyone's responsibility to aim for quality by doing things right the first time. This would not only reduce the amount of scraps and rejects from reworks but help the firm win happy customers. Ultimately, the goal of any firm should be to satisfy and maintain loyal customers. Lean manufacturing cannot foster if total quality control is nonexistent. Just like the JIT system, TQC is a strategy to minimize waste and that is a prime focus of "lean" manufacturing.

## Total Productive Maintenance

The lifeblood of any manufacturing system is the process of transforming inputs into outputs or finished products. Many of the manufacturing processes today are highly mechanized and any malfunction or even inability to meet tolerance requirements would create large amounts of waste. Equipment failures can be traced to a range of causes including poor equipment design and human errors. The aim of total productive maintenance is to ensure that these problems do not occur by adopting strategies that would help minimize equipment downtime. Such strategies may include scheduled preventive maintenance programs, enhanced operator training programs, maintaining equipment backup or standbys in case of unpredicted failures, etc. However, whatever strategy is adopted should be cognizant of the type and distribution of equipment failures as well as its associated costs. Again, lean manufacturing cannot be achieved if the manufacturing process is unpredictable and incapable of performing within established standards and guidelines.

## Flow Manufacturing

Flow manufacturing is a hybrid manufacturing strategy that exploits the strengths of other manufacturing techniques such as agile manufacturing, lean manufacturing, synchronous manufacturing, just-in-time manufacturing, and demand flow technology. It is based on a complete re-engineering of the entire manufacturing system rather than trying to achieve incremental gains or improvements. This drastic and radical approach to manufacturing relies on the premise that in order to compete in today's dynamic environment, a new manufacturing style that is proactive, flexible and dynamic must be applied. Baum [2001] notes "flow manufacturing provides the flexibility of mass customization environments with the efficiency of the classic assembly line." He further states that many of the traditional manufacturing practices such as material requirements planning (MRP), work order routings are not essential in flow manufacturing. In fact, MRP is used in flow manufacturing only to plan long-term needs and maintain vendor blanket orders and relationships. It is not needed for capacity requirements planning since flow manufacturing is based on a balanced production line.

Like the other manufacturing systems we have discussed in this chapter especially the JIT, flow manufacturing is a pull-driven system that is driven by customer orders. This contradicts the push-driven system used in the traditional assembly line operation. Furthermore, the flow manufacturing operation utilizes the good qualities of an assembly line system by ensuring smooth flows of work through the line but deviates from assembly line systems by maintaining product flexibility and using the pull system to minimize the inventory requirements. This system does not maintain work in process (WIP) inventories and ensures that high quality is maintained at a minimum cost. Flow manufacturing aims to derive maximum value from an activity while expending minimum efforts and energy. It is a value-based strategy that aims at eliminating waste and adding more value to the manufacturing operation.

## Conclusion

This chapter's focus is on manufacturing strategy and we have specifically discussed the roles of agile, lean, and flow manufacturing in achieving competitiveness in our dynamic and highly competitive environment and in managing environmental wastes. They result in increased productivity as more value is obtained from less input. Thus, vital and limited natural resources in the form of materials and energy are conserved. In addition, when quality is high, the number of scraps, rejects, and reworks is reduced thereby placing less demand on the need for more resources. The number of production runs is also less thus demanding less energy resources. The ripple effect of all these is that environmental burden is minimized. Lesser materials and wastes will be targeted for disposal at the landfills and there will be less need to excavate and exploit new natural resources. Efficient manufacturing systems can therefore, significantly help to achieve not only the quality imperative of a firm but also, the environmental needs of the society. Waste reductions in inventory or by doing things right the first time, conserve energy, materials and other resources. Thus these strategies are environmentally conscious. These manufacturing strategies focus on achieving high quality and productivity by minimizing waste.

These new strategies are gradually replacing the traditional mass production system that is push- rather than pull-driven. The push-driven system encourages large volume production, limited flexibility, quantity against quality, and large inventories. It is not robust and responsive to the dynamic changes in the environment. It encourages waste creation and it is not sensitive to environmental needs. Given that our environment is unpredictable and very dynamic, new manufacturing strategies that are proactive and flexible to respond to uncertainties and unpredictability in the environment are in dire need. Organizations must transform themselves by adapting to these new manufacturing strategies if they intend to remain competitive and capable of satisfying the changing needs and demands of their customers especially as they relate to the natural environment.

The new manufacturing technologies discussed in this chapter offer effective means to meet the environmental needs of the firm and the society. Their application should be integrated in the framework of other environmentally conscious manufacturing strategies that have already been discussed in this book.

## References

Agile Enterprise/Next Generation Manufacturing Enterprise,"
  http://www.CheshireHenbury.com, downloaded September 9, 2001.
"Agile manufacturing,"
  http://www.technet.pnl.gov/dme/agile/index.htm,
  downloaded September 9, 2001.
Baudin, M., "The meaning of "lean."" Dated 8/11/97
  http://www.mmt-inst.com/Meaning_of_lean.htm,
  downloaded September 26, 2001.
Baum, D., "Flow manufacturing,"
  http://www.oracle.com/oramag/profit/98-May/flow.htm,
  downloaded September 26, 2001.
Keen, P.G.W., "Agile manufacturing,"
  http://www.peterkeen.com/emgbp003.htm,
  downloaded September 9, 2001.
"Lean manufacturing," http://www.cimplest.com/leanmfg.html,
  downloaded September 20, 2001.
Maskel, B.H., "An introduction to agile manufacturing,"
  http://www.maskel.com, downloaded September 20, 2001.

**Chapter 10**

# Environmental Risk Assessment and Management

Risk assessment is a major component of environmental management. Risk is a measure of the likelihood or the probability that an event will occur. Risk is used in almost any situation where there are uncertainties. The aim of risk measures is to be able to estimate the likelihood of an occurrence in order to plan better or make better decisions. Environmental hazards are of major concern in managing environment. Such hazards may lead to destruction of lives and properties, destruction of wild life, long-term illnesses and diseases, pollution of the natural environment, and high clean up or litigation costs. In order to plan properly, it is important to estimate the risk or rather, the probability of an environmental hazard. Such risk assessments may include for example, the environmental risk of an oil spillage, hazardous and toxic waste dumps, genetically modified organisms, nuclear power plant disasters. The aim is therefore, to estimate the potential that the occurrence of any of these events could cause harm. Thus, *hazard* implies the possibility of harm to people or the natural environment.

Hazard and risks are frequently referred to in environmental risk assessment. They two are not the same. Hazard as we have just defined, is the potential for harm while *risk* is the degree or the likelihood of harm. For example, exposure to ultra violent rays from the sun could be classified as a hazard to human health. However, there are different degrees of exposure which are measured by risk. Since certain hazard may be inevitable, there would always be risks involved. The degree of risks would determine an *acceptable* or *tolerable* risk in any given situation.

The aim of risk assessment is to estimate the risks that are posed by the hazards that are inherent in a particular process. For example, all production processes create some form of environmental hazards. They all emit gases, expend natural resources, and consume energy. However, we classify some of these processes as *cleaner* and not necessarily *clean*. This underscores the fact that there are some potential for hazards in the process. However, the "cleaner" process may pose acceptable or tolerable risks to humans and the natural environment. Thus, the intent is not to achieve a *zero risk* but rather, to minimize the risk level to an acceptable level of risk. Such an acceptable risk would imply low probability of environmental or health hazard. It is important to note that in considering potential hazard, it is also necessary to consider the severity of the hazard in the computation of the risk.

Environmental risk assessment (ERA) is comprised of Health Risk Assessment and Ecological Risk Assessment. The former is a measure of the risk of hazards on humans while the later measures the risk of hazards on the ecosystem. In a report by the European Environmental Agency[15], it defines Environmental Risk Assessment (ERA) as "The examination of risks resulting from technology, which threaten ecosystems, animals and people. It includes human health risk assessments, ecological or ecotoxicological risk assessments, and specific industrial applications of risk assessment that examine end-points in people, biota or ecosystems."

Risk management is scenario-driven. The reason risk assessment is important is because of the need to make better decisions. Thus, in environmental planning, different scenarios are investigated and their risks assessed to identify an option with tolerable risks. This assessment presumes that accurate information or data about potential hazards are obtained. However, this is not always the case. With the rapid proliferation of new products and technologies, we are often faced with incomplete or insufficient data about these new products. As such, some of the environmental hazards may not be adequately estimated or even anticipated.

The acceptable risk option becomes a logical selection for implementation. Usually, such risk assessment is guided by established

---

[15] "Environmental Risk Assessment – Approaches, Experiences and Information," Environmental Issue Report No. 4., http://reports.eea.eu.int/GH-07-97-595-EN-C2/en/iss/inth.html, downloaded March 19, 2005.

environmental standards and laws and regulations. The firm sells the idea of acceptable risk to its stakeholders. Risk acceptance by the stakeholder is however, affected by risk perceptions. The worldviews, culture and value systems, and the beliefs of the stakeholders play major roles in determining their acceptable risks. Their perceptions would influence greatly how risks are managed. We also note that risk acceptance and management are influenced by knowledge and information available to the stakeholder. Often times, as in many developing economies, stakeholders may not have the information or knowledge to properly evaluate their risks. Therefore, unacceptable risks may be taken due to lack of knowledge or information. An example is the excessive use of pesticides and synthetic fertilizer products by subsistence farmers in developing countries. Due to lack of adequate training and information, many of the farmers expose themselves to hazardous substances and, deplete their farm land of organic matters that foster higher farm yields. Yet, because they are often unaware of the consequences of the application of these chemical compounds, they are accepting unusually higher levels of risks.

## Effective Use of Risk Assessment

The essence of risk assessment is the ability to make informal decisions and better policies. Risk assessments are not only used by corporations but also by policy makers. We identify some of the key areas where risk assessments are applied:

- Environmental regulation – Many government environmental protection agencies rely on risk assessment to determine societal acceptable risk levels. Based on risk assessments, regulations on natural environment are established. Standards for product environmental content, waste disposal, emissions, toxic and hazardous waste management are established.
- Land-use laws often require environmental impact assessment. Such assessments usually would conduct risk assessment to study the risk of certain development projects on wild life, traffic

congestion, crowding, urban migration, and underground water supply. For example, location of gulf course is often challenged by communities because of the excessive use of pesticides and fertilizer to keep it green. Also, when large apartment complexes are proposed, surrounding communities worry about increased pollution and poor quality of life in terms of noise, traffic, waste, etc. Additionally, social problems may be associated with increased population. Thus, the impacts of such development efforts on the well-being of the town and on the environment are investigated using a risk assessment model.

- Regulation of chemicals and other hazardous waste – Since it is not possible to achieve zero risk, chemical substances and other hazardous wastes should be prioritized based on their risk levels. For example, the banning of chemical compounds such as CFC, DDT reflects the highly unacceptable risk associated with them. Risk assessment also helps to regulate how certain materials are used. For example, the collection and proper recycling of used motor or engine oil reflects the need to limit spillage and unsafe disposal of such products. Also, measures to control the disposal of medical syringes because of health hazards show again, how efforts are made to control risks.

- Substitute – The high risk associated with products in terms of health and ecological risks may necessitate the need to find alternative or substitute products. For example, the fast food chain McDonald's switched from polystyrene containers to paper wraps for its packaging thereby reducing the amount of solid wastes it creates. The use of recycle paper products as against paper derived from 100% virgin pulp has significantly reduced the demand for cutting down forestry.

- Industry standards – Different industries establish their ecological and product safety standards to complement public efforts. They engage in research and development to find alternative and eco-friendly designs, processes, and substitutes to minimize or reduce current ecological and health risks.

## How is Risk Assessment Affected?

Improve design – When risk is high, the potential for harms to humans or ecology is high. The government comes in to protect the public. An alarm about the product or process is created possibly leading to plant closures. For example, the Nuclear Regulatory Commission (NRC) frequently monitors and regulates nuclear power plants to assess their risks. High risks pose danger to communities and exceed the bound of "acceptable" risk. To protect the public, the plant may be closed down until proper measures are taken to lower the risk.

Companies that face the possibility of shutting down or withdrawing their product from the market because of the unacceptable risk associated with the products; face the potential of losing market share to their competitors. Rather than taking corrective measures, it is better to adopt preventive measures in risk assessment. In 2004, the popular painkiller by Merck & Co. known as Viox was withdrawn from the market because of its link to heart diseases. The stock price for Merck sunk by over 50% and on August 19, 2005, a Texas jury awarded $253.4 million in damages against Merck & Co to a widow of a man who had a sudden cardiac death[16]. Apparently, the man died of arrhythmia – a type of heart problem that was not then linked to Viox. However, the fact that Viox was associated with the risk of heart disease was enough to convince the jury. There are also thousands of lawsuits against Merck & Co in other states because of this health hazard and potential risks associated with its product Viox. Companies could therefore limit their exposure to risks by designing their products or processes to manage hazards. Such design strategy would involve life cycle assessments to evaluate different options and investigate their environmental impact. An option with lower and acceptable risk level is accepted and designed in to the product or process.

New technologies – New technologies are generally, "cleaner" because they are designed for the environment. Their designs also pay attention to new and evolving international and national environmental

---

[16] "Merck Jerked,"
http://www.reason.com/hitandrun/2005/08/merck_jerked.shtml

laws and standards. There is a need to continue to modify, and/or adopt new technologies that meet established environmental performance standards. By meeting such standards, environmental performance is enhanced and the risk level is minimized.

Selection of new technologies to reach acceptable risk should be based on established environmental guidelines and standards. In fact, industry standards exist to complement international and national standards and should serve as guide to attaining environmental performance. ISO 14000 series of Environmental Management Systems devote considerable amount of effort on Environmental Performance Evaluation[17]. ISO 14031 focused on Environmental Performance Evaluation (EPE) which it defines as "A process to select environmental indicators and to measure, analyze, assess, report and communicate an organization's environmental performance against its environmental performance criteria." There are several environmental performance indicators (EPI) that are used to check compliance to environmental performance criteria. However, the criteria are based on acceptable risk as a guide and the aim is to meet or exceed such established standards. Companies that exceed the standard achieve competitive advantage and could become benchmark companies. Companies like Toyota Motor Company and Honda that were among the early leaders in introducing hybrid vehicles that are not only environmentally friendly but also economical, are gaining huge market share and are increasingly competitive in the highly competitive auto industry.

Major environmental performance indicators that companies could use for their risk assessment include the following:

- Resource consumption
- Water requirements including waste and polluted water
- Toxic wastes
- Greenhouse gases
- Waste disposals
- Packaging requirements
- Distribution and logistics requirements

---

[17] "Introduction to ISO" ISO Online, http://www.iso.ch/infoe/guide.html.

- Urban pollution (noise, smug, odor)
- Air pollution
- Loss of biodiversity
- Energy requirements

These indicators cover some aspects of pollution and resource depletion that could create ecological and health hazards. Proper accounting or inventory for these indicators would help to effectively estimate the risks associated with the product or process. Risk assessment should look at several environmental performance indicators and not just a single indicator. Many companies today, document in their annual environmental report, their contributions to environmental burden. Such reports often referred to as *green accounting*, increasingly help to give a better view of how the company contributes through its operations and activities in achieving environmentally conscious manufacturing.

Figure 10.1 identifies the different types of risk assessment and issues involved in the assessment of risks. These risk areas are not completely independent since a particular hazard could affect all types of risks. In ecological risk assessment, due attention is given to the influence of environmental burdens on different environmental media such as air, water and land. The consequences of such hazards such as land-use, waste and chemical deposits, emission of gases, and resource depletion are estimated. With health risk assessment, attention shifts to the health of the general public. Possible consequences of environmental burden on human health is investigated and assessed as they would affect the quality of life, productivity, and short- and long-term health care management. Industrial risk assessment investigates environmental burden by focusing on industrial management of waste, cleanup litigation costs, process and product changes, customer perceptions, logistics and distributions, and also the role of the supply chain. These different views of risk help improve the estimation of the risks involved and help to guide effective policy and decision making. From Figure 10.1, it is clear that there are several sources of environmental hazards and it would be very difficult to fully account for all of them and examine all potential environmental hazards.

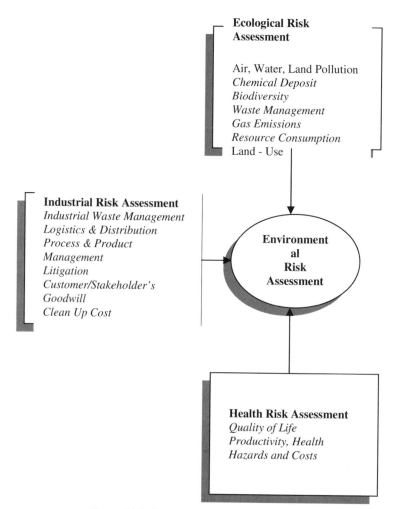

*Figure 10.1: Environmental Risk Assessment*

In 1983, the US National Research Council (NRC)[18] introduced one of the most widely used methods to assess the risk of chemicals on human health. This model is most widely used by regulatory agencies in conducting health risk assessment.

---

[18]   NRC 1983 <u>Risk Assessment in the Federal Government: Managing the Process</u>. US National Research Council, National Academy Press, Washington DC.

According to Covello and Merkhofer[19], all risk assessment models must consist of the following steps:

- Problem formulation
- Hazard identification
- Release assessment
- Exposure assessment
- Consequence assessment
- Risk estimation

We shall briefly describe these steps.

Problem formulation – This is the first step in any problem solving. It is important to define the boundary and the scope of the problem. With a good problem formulation, it is easier to narrow the problem solving down so that the critical aspects of the problem are investigated. This stage also helps to understand the objective of the problem. Critical issues addressed include the reasons why the problem is being solved; and the variables and constraints involved.

Hazard identification/Release assessment – This approach is based on the use of tools to understand potential hazards and the likelihood of a release. There are different tools that could be used here. Some of them are fault tree/event tree analysis, failure mode and effect analysis (FMEA), hazard indices, and cause-and-effect diagram. Some of these tools are used in other fields of study but have been found pertinent in studying environmental risks. We have discussed the cause-and-effect diagram in other parts of this book. We shall discuss the other important tools that we have identified.

Fault tree/event tree analysis (ETA) is a means of analyzing causes or sequences of possible events that may occur in a system. It is a visual representation of all possible outcomes. It is often used to determine the probability of an event based on the outcomes of each event. Thus, the probability that an outcome may lead to a desired result is estimated.

---

[19] Covello, V.T. and Merkhofer, M.W., 1993, <u>Risk Assessment Methods: Approaches for Assessing Health and Environmental Risks:</u> Boston, MA: Kluwer Academic Publishers.

FMEA is another powerful tool. Basically, FMEA is used to identify the effects or consequences of a potential product or process failure. It offers an approach to eliminate or reduce the likelihood or occurrence of failures.

Hazard indices are used to assess human exposure to environmental pollutants. Hazard index is defined as[20]

$$Q/ Q / sub L/$$

Where

$Q$ = exposure or dose to total-body, organ or tissue from all environmental pathways; and

$Q / sub L / $ = a limit that should not be exceeded because of health risk to humans.

There are different hazard indices for the different sampling medium and they correspond to each effluent type. A composite index is obtained that ensures that health risk limit is not exceeded.

Exposure Assessment – Here, assessment is made of the magnitude of the physical effects of hazard, as well as its pathway or transportation mode to the receptor. There are different models to predict exposure to radiation, heat, sound as a result of explosion, or spray of chemicals or pesticides. For example, in the US, the Office of Pollution Prevention and Toxics (OPPT) has developed several exposure assessment models[21]. These predictive models are used to evaluate:

- What happens to chemicals when they are used and released to the environment; and
- How workers, the general public, consumers and the aquatic ecosystems may be exposed to chemicals.

---

[20] Walsh, P.J., Killough, G.G., Parzyck, D.C., et al., "CUMEX: a cumulative hazard index for assessing limiting exposures to environmental pollutants," Report No. 7224379, Ap 01, 1977, http://www.osti.gov/energycitations/product.biblio.jsp?osti_id=7224379.

[21] "Exposure Assessment Tools and Models," http://www.epa.gov/opptintr/exposure/.

These models may be helpful when monitoring of data may not be available or may be insufficient. These models could help in the consideration of potential exposure in the designing and selection of products and processes. They would also help in the evaluation of pollution prevention opportunities.

Consequence Assessment – The consequences of the exposure to hazard are quantified. This assessment estimates the damage to the receptor due to the exposure. Vulnerability models are used at this stage. Many of the vulnerability models focus on physical and structural vulnerability of natural hazards. However, the social and psychological elements are often overlooked[22]. It is also difficult to attribute the effects associated with hazards to any one cause for an event. Thus, a holistic view of hazards and their causes should be adopted.

Risk Assessment – An estimate of the overall risk associated with an activity is obtained by integrating the probability estimates of release events and the consequence assessment. This integration process may involve comparing single values of effect and exposure. Statistical distributions of exposure and effect values could also be compared. Often times, simulation models may be developed to evaluate alternative scenarios.

Fairman, Mead, and William[23] present a flow diagram to represent environmental risk assessment. We shall adapt this diagram and present it below.

---

[22]  Jones, B., "Vulnerability Models," http://www.utoronto.ca/env/nato/proceedings-October 2003/ discussion1.pdf.

[23]  Fairman, R., Mead, C.D., and William, WP, 1998, "Environmental Risk Assessment: Approaches, Experiences and Information Sources," Environmental Issues Series No. 4., Copenhagen European Environmental Agency.

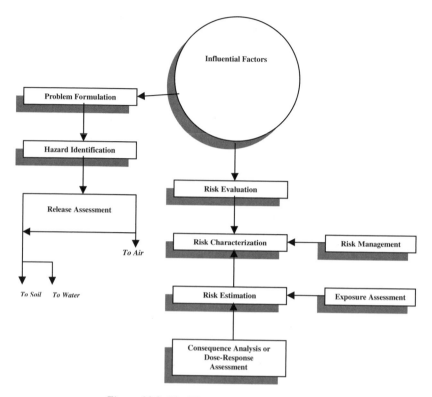

*Figure 10.2: The Elements of Risk Assessment*

Figure 10.2 basically, summarizes the discussion on risk assessment. It is based on the steps for risk assessment models that were discussed above. The framework starts with problem formulation. This step defines the existing problem and the reasons to assess its potential impacts. The problem formulation is however, influenced by a systemic view of the natural environment which includes economic, political, social and legal factors. All of these play a role in problem formulation and definition. Many nations resisted attempts to cut down on carbon emissions since they perceive an association between carbon emission and economic growth. Yet, this singular issue has an overreaching influence on economic development, social, political, and legal factors. The potential hazards from an activity are identified and the releases are assessed on the different environmental media – land, air and water.

Risk evaluation is also performed to determine the significance of the risk to those that are affected, those that create it, and those who control it. However, there are issues that affect effective evaluation of risks. As we have already mentioned in this chapter, it is not always easy to evaluate potential risks. Certain important information may be unknown at the time of the evaluation. So, the importance of the risk to the receptor may not be fully understood and explained. The goal of risk management should therefore, be to characterize and attempt to minimize the risk. A quantitative or qualitative measure of risk is then conducted by combining exposure assessment and consequence analysis.

## Environmental Action Box
## Case Study on Bristol-Myers Squibb

In this chapter, we present the case of Bristol-Myers Squibb – the pharmaceutical drug company and the strategies it has taken to improve environmental performance and health safety. These strategies have helped the company to minimize ecological and health risks. We also demonstrate the perception of risks by presenting the case of GE and Housatonic River.

Bristol-Myers Squibb has a long history of voluntary participation in environmental, health and safety (EHS) program. It has endorsed many of the worldwide efforts by International Chamber of Commerce (ICC) Business Charter for Sustainable Development, and the US Environmental Protection Agency programs to reduce releases of toxic chemicals[24]. It embarks on a pollution prevention program that tracks product life cycle and views environmental management among its corporate priorities in its strategic plan.

In its worldwide operations, it engages customers and suppliers in its environmental management system programs by providing them

---

[24] "Our Environmental Building Upon Our Successes," Office of Environment, Health and Safety, NY, NY. http://www.bms.com/ehs.

with safety and environmental information. Its customers are mainly hospitals and health professionals. Its efforts include helping hospitals to develop ISO 14001-type environmental standards. According to its Europe regional vice president, Ake Wilkstrom, "Our environmental, health and safety efforts are wonderful examples of how we can apply in-house competencies to reduce *risks* and provide a valuable service to our customers at a very low cost. Ultimately, it helps to improve our image and drive sales.[25]"

Bristol-Myers Squibb achieves its environmental goals by designing for the environment. Its research and manufacturing teams focus on how to lessen the environmental impacts or *hazards* of materials used to produce its goods or products. One of its major products is the *Keri* line of skin care products. The product has been re-designed for the environment by replacing 50% of the emulsifiers in Keri Fast Absorbing Dry Skin Formula with stearic acid that is naturally derived and biodegradable. Further, propylene glycol which is nonrenewable resource has been replaced in the product with glycerin which is both plant-derived and environmentally safe. Other product lines such as Clairol hair sprays have also been transformed to ensure environmental sensitivity. These developments have also extended to cardiovascular therapies where the use of volatile solvents have either been reduced or eliminated.

Bristol-Myers Squibb's credits its efforts with lower accident risks at its plants. For example, in 1995, recorded injuries at its Stamford, CT plant for Clairol were 123. By 1999, this number has dropped to 53.

Bristol-Myers Squibb environmental program is based on the estimation of net value by looking at the integration of economic, social and environmental impact on society. An adapted diagram is shown in Figure 10.3 below:

---

[25]  ibid

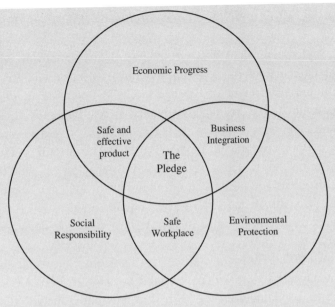

*Figure 10.3: Integrating Economic, Social and Environmental Issues*

The goal of Bristol-Myers Squibb's environmental program is to increase or maximize shareholders' wealth and increase value to the community while reducing environmental burden and hazard. Employees play major role in this since their involvement is important to achieving this goal. Likewise, suppliers are active participants in this effort. Environmental impacts on all the environmental media (land, air, water) as well as on natural resources, biodiversity, and human health are assessed. The sustainability program is also followed in Bristol-Myers Squibb's packaging program. Significant reductions have been achieved in cardboard usage and the volume of packaging materials handled by the customer. Recycled papers are used and metal contents in its cans for infant formula have been significantly reduced. Products are designed to achieve further reduction in size and reduce the demand for packaging materials.

## *General Electric (GE) and the pollution of Housatonic River*

On September 24, 1998, GE endorsed an agreement with the State of Connecticut to clean up and restore Housatonic River that was polluted from General Electric's (GE's) operation in Pittsfield, Massachusetts[26]. The cleanup is estimated to cost hundreds of millions of dollars. The contamination of Housatonic River is a result of the release of polychlorinated biphenyls (PCBs) from GE's Pittsfield facility to the river. These releases have impacted on wildlife and natural resources and present hazard to human health. The future of the river is also threatened. PCBs were banned since 1977 and are considered as probable cause of cancer in people and developmental disabilities in children.

In April 2004, Housatonic River was named as one of the nation's Most Endangered Rivers with concern that the slow effort by GE to clean up toxic PCBs dumped in the river threatens public health along the length of the river[27]. This concern was partly based on the long history of GE dumping PCBs from its electric transformer manufacturing plant in the river between 1932 and 1977 and its aggressive challenge of EPA's demand for cleanup in court[28]. The river is now said to be contaminated with PCBs from the river bottom all the way to its mouth in Long Island Sound. Nearby lakes have also been affected. The pollution has entered the food chain through plants and bottom-feeding species. Housatonic River was ranked 7th in the list of America Most Endangered Rivers by www.americanrivers.org. The levels of PCB found on ducks from the most polluted section of the river were more than 200 times EPA's tolerance level for human consumption. The carcasses of these birds must therefore be handled as hazardous waste. Fish consumption advisory was placed as early as 1977 stretching about 100 miles of

---

[26] "Connecticut supports agreement with GE," Connecticut Department of Environmental Protection, http://dep.state.ct.us/whatshap/press/1998/mf092498.htm, September 24. 1998.

[27] "Housatonic River #7 on annual list released today," http://www.americanrivers.org, April 14, 2004.

[28] "Housatonic River "Most Endangered"" http://www.ems.org, April 14, 2004.

the river from Pittsfield, MA. Consumption of contaminated fish or waterfowl could lead to cancer or reproductive problems. The states largely affected by this contamination are Connecticut and Massachusetts.

In the agreement reached with the state of Connecticut, GE was obligated to perform the following:

Pay a compensation package for natural resource damages that is valued at $25 million and to the remediation of a two-mile section of the Housatonic River heavily contaminated by polychlorinated biphenyls (PCBs). The States of Connecticut and Massachusetts will share the $25 million dollars. GE is also to carry out environmental enhancement and restoration projects in and around the City of Pittsfield.

In March 2005, the US Environmental Protection Agency proposed to issue a permit to GE that would allow it to discharge PCB-contaminated water into the Housatonic River[29]. This has put the EPA and GE at loggerheads with community activist groups who oppose the proposal and view it as a reward to GE's bad environmental practice while increasing the ecological and health risks nature, wildlife and humans face as a result of this continuing act of pollution. This discharge permit is being issued jointly by the US EPA and the State of Connecticut Department of Environmental Protection under the National Pollutant Discharge Elimination System (NPDES).

This report shows that there is a high perception of risk associated with further releases of PCBs to Housatonic River. This perception of risk has led to resistance of the policies of GE and the actions of the environmental protection agencies. Such perceptions may place GE in conflict with its community or operating environment. This may lead to negative image that may not support GE's organizational goal and social responsibility function.

---

[29] Eagle, B., "Activists say EPA lets GE pollute river," *Connecticut Post*, March 25, 2005, p. D5.

## Conclusion

In this chapter, we focused on the basic introduction of risk assessment. In environmental management, it is important to quantify the risks associated with hazards that result from environmental burden. We focused mostly on ecological and health risk assessment. These risks contribute significantly to the public reactions on the degradation of the natural environment.

We discussed the stages of risk assessment and we presented frameworks and some important models for these stages. Risk assessment is complex and often requires a strong foundation on probability theory and statistics. Such background is beyond the scope of this book. However, interested readers may find specialized books on this topic.

We conclude by noting that environmental risks could be minimized by taking some precautionary measures. Such measures include the following:

- Designing for the environment – Throughout this book, we have discussed DfE. The use of DfE enables consideration of the total production system to ensure that the product and process are designed to minimize the creation of environmental burden or hazards to the natural environment. This could be affected through effective product, process or vendor selection, use of substitute material i.e., the use of organic rather than inorganic fertilizer, use of recycled paper, etc., design for reuse, and other strategies that would lead to environmentally conscious manufacturing.

- Recycling policies that provide detailed information on how products can be safely disposed off could significantly help to minimize hazards and lower the associated risks. Several industries maintain effective recycling programs and incentives to encourage participation in the programs.

- An educated consumer can significantly lower his risks by evaluating the environmental contents of the product. With the availability of virtually all information on the Internet, an educated consumer is better guided to make decisions that would lower his risks.

- Legislatures and laws that impose heavy fines and penalties on polluters also help the public by limiting the production of hazardous products and establishing acceptable standards for products."
- Finally, with many nations adopting ISO standards, ISO 14000 could present a universal guide to Environmental Management System thereby reducing the environmental risks to the public.

An environmental action box is presented to show actions taken by Bristol-Myers Squibb to reduce its environmental impact and thereby minimize ecological and health risks. We also showed how risk perceptions could lead to resistance of ecological policies. This is shown with the case of GE and Housatonic River.

# Chapter 11

# Competing on Environmental Management

This chapter summarizes the book by focusing on the management of environment as an important competitive weapon. Specifically, the role of environmental strategies in satisfying stakeholders' needs is investigated further. Managing environmental burden is viewed as a social responsibility function and may influence not only the competitiveness but the survival of the firm.

There is a growing awareness of the importance of protecting the natural environment through sustainable or environmentally conscious manufacturing practices. The world community is concerned about the quality of air, water, and land. Most people want to trim waste by minimizing solid and hazardous waste disposal, reducing the emission of poisonous and toxic gases to the atmosphere, controlling the emission of greenhouse gases that destroy the ozone layer, and maintaining ecological balance by re-evaluating excavation and development policies. The need to protect the earth's limited natural resources is important and was necessary to convene two international conferences the Earth Summit held in Rio de Janeiro, Brazil in 1992, and the Global Climate conference held in Kyoto, Japan in 1997. The world business community responded to the public demand to protect the environment and conserve natural resources through the Brundtland Report of July 1987, which challenged businesses to achieve sustainable economic growth without compromising the natural environment. Many nations around the world adopted strict measures on the environment making it difficult to conduct business as usual without a focus on environmental protection. In the US, the Clean Air Act was amended in 1990 with more stringent laws and penalties for pollution. Strict laws were enacted on the use of ozone depletants such as CFCs (chlorofluorocarbons), methyl chloroform, and carbon tetrachloride, all which were outlawed in the year 2002. Similar laws were also passed in countries such as Germany and Holland. These laws do not only outline penalties for pollutants but also set guidelines and procedures to achieve environmental goals. Some of the environmental strategies encouraged include:

- The use of recyclable materials
- Designing products for ease of assembly and dismantle
- Use of reverse logistics to reclaim old equipment from users and properly disassemble them to recycle and remanufacture useful parts
- Creation of new applications for used materials
- Safe disposal of unusable materials and equipment.

These requirements created new challenges and opportunities that businesses must deal with. Today's organization must compete on the basis of its environmental practice.

## The Social Responsibility Function

The responsibilities of corporations as business entities interested in maximizing profits and shareholders' wealth have changed. Corporations today also have to be good corporate citizens. They should be exemplary in their communities and have diverse roles to play. The business of the corporation is the business of its community and the society at large. It is not only the business of its shareholders or its direct customers. Therefore, more is expected of corporations today. They do not only provide specialized products and services but are expected to form long-term partnership with their communities. Corporate missions must therefore be aligned with the long-term goals, missions, and vision of the community. While shareholders may be interested in maximizing their wealth, members of the community are more interested in maximizing their quality of life. These goals can only be achieved by working with the corporation to ensure that corporate goals, policies, and strategies are in line with community goals and expectations. Thus in order to achieve the social responsibility goals, business organizations must have shared mission, vision and goals with their communities or operating environment. Gone are the days when the main social responsibility function for a corporation was to provide jobs to its community. Members of the community are now more enlightened and look beyond the short-term values that a firm may offer but seek to develop a sustainable partnership that cannot only help them but also their children and the future of their natural environment. In the absence of a vision that is shared by the community, the firm stands to face increased opposition from its community, which could significantly harm the image of the organization and its ability to conduct business operations. Social responsibility can only be satisfied when the business is ethical. We have noticed in recent times, public condemnation of companies that have been involved in ecological mishaps or disorders. For example, the Exxon Valdez oil spill and the Union Carbide plant explosion in Bhopal, India not only brought litigation and fines to these companies but also public condemnation and discontent which harmed the image of both companies. There is therefore, a greater need to focus on environmental protection as a way of achieving the social responsibility of the organization. Corporations need to start working with "stakeholders" rather than just stockholders. Stakeholders are active participants who are affected by the decisions and activities of the firm or whose activities and actions may affect the future operations of the firm. In today's litigious environment, stakeholders who feel that the quality of their lives have been compromised by the actions of firms are increasingly suing corporations and courts are imposing significant penalties for corporate malfeasance. Businesses must therefore listen to the voice of the stakeholders.

## Product Stewardship

The Northwest Product Stewardship Council (NPSC) defined product stewardship as follows [Bullard, 1994]:

"Product stewardship is a principle that directs all actors in the life cycle of a product to minimize the impact of the product on the environment. The concept is unique because of its emphasis on the entire product system. Under product stewardship, all participants in the product life cycle—designers, suppliers, manufacturers, distributors, retailers, consumers, recyclers and disposers—share responsibility for the environmental effects of products."

Companies are responsible for tracking the environmental burden of their products through their life cycles. This call for responsibility has created huge burden to corporations that have not paid attention to the environmental content of their products. There is no time limit to the company's responsibilities when it comes to environmental burden.

Polluters are increasingly forced to take responsibility for their acts. One of the most devastating environmental pollution of the 20th century in the United States was the pollution of the Love Canal. This was a case of natural disaster of untold and unimaginable human proportion. Between 1942 and 1953, Hooker Chemical Company now part of Occidental Petroleum and Chemical Corporation dumped 20,000 to 25,000 tons of toxic chemicals into the Love Canal. Many of the chemicals dumped were pesticide waste and chemical weapons research (i.e. The Manhattan Project) [Allan, 1998] and are listed in the order of largest concentrations as benzene hexachloride, chlorobenzenes, and dodecyl mercaptan. Although much may be known about the health effects of a single chemical, little is known about exposures to a mix of synthetic chemicals. In Love Canal, more than 200 chemicals and toxics were disposed and absorbed by the soil. The impact was quite devastating as dangerous chemicals such as dioxin and mercury seeped through the soil and polluted the entire area. Studies showed that women living in the area were having higher rates of miscarriages, stillbirths, crib deaths, and childhood neurological problems and hyperactivity [Gibbs, 1999]. Bullard [1994] attributed the complacency of companies like Hooker Chemical then to the environmental policies that focused on how to manage, regulate, and distribute risks. That led to a dominant environmental protection paradigm that was according to Bullard, based on the following principles:

- Institutionalizes unequal enforcement
- Trades human health for profit
- Places the burden of proof on the "victim," not on the polluting industry
- Legitimizes human exposure to harmful chemicals, pesticides, and hazardous substances
- Promotes "risky" technologies
- Exploits the vulnerability of economically and politically disenfranchised communities

- Subsidizes ecological destruction
- Creates an industry around risk assessment
- Delays cleanup actions
- Fails to develop pollution prevention as the overarching and dominant strategy

History can buttress Bullard's points as we note that many industries and corporations then were anti-pollution prevention and sought end-of-pipe management of pollution problems. They adopted the "quick fix" strategy to environmental management and would then respond only when a problem occurs. For example, some of the companies that are being heralded today for championing the environmental movement were once at loggerheads with environmental protection strategies. The Exxon-Valdez Oil Spill was as a result of the single-hulled oil tanker that ran aground and spilled 11 million gallons of oil into the Prince William Sound, Alaska. Prior to this disaster, the oil industry fought against legislation that required the use of double-hulled oil tankers. Likewise, the fossil fuel industry used to work hard to discredit efforts on global warming by claiming it was a myth. Chemical companies such as DuPont took a long time before acknowledging that CFCs were destroying the environment, and McDonald's even filed a libel suit against London Greenpeace for criticizing McDonald's role in rainforest destruction and other practices.

The environment has however changed since Bullard made his assertions. More and more corporations are being made to take full responsibility for their acts irrespective of when it occurred and what information they had at the time. The environment today is more litigious than twenty years ago. Further, many states and countries have responded to public outcries and enacted new laws to ensure corporate responsibility towards the environment and to protect the rights of citizens and communities. These laws have made companies vulnerable to lawsuits from stakeholders and they are responding by taking a product stewardship approach for their products.

Product stewardship requires taking a cradle-to-grave approach of one's products and services. It starts from product design and involves all stages of product development, distribution, consumption, and disposal.

This therefore, requires that the manufacturer work with its suppliers and vendors to ensure that this goal is achieved. Madu [1996] reported that companies and industries in Japan are already ahead in environmental friendly practices as the next competitive weapon. This role is also fostered by some of their manufacturing practices such as the Just-in-time and lean manufacturing practices that are inadvertently, supportive of the goal for environmentally conscious manufacturing. Madu notes that industries ranging from auto, steel, heavy metals, and energy in Japan have already adopted environmentally friendly practices to help them compete in the new millennium. American companies are also heeding the call to be environmentally conscious. Many corporations are now integrating environmentally friendly practices in their mission statements and are assigning strategic importance to them. We have given some examples of corporate actions and practices in the Environmental Action Boxes and would like to refer the reader to those case examples. Here, we emphasize on how suppliers can help support the environmental practices of a firm.

## Supplier Participation in Environmental Practice

Companies today are adjusting their strategies of selecting suppliers to move away from a low bid approach and look at a whole range of other factors that may influence quality, productivity, effectiveness, and customer satisfaction. Although supplier selection factors may include issues such as quality, reliability, cost, flexibility of the supplier, a critical element that is now frequently considered is the environmental practice of the supplier. Manufacturers look beyond the bidding system to develop a long-term partnership with a supplier and this long-term outlook requires that the manufacturer also evaluates the long-term costs that may be absorbed by partnering with the supplier. Such long-term costs may be related to the environmental practice of the supplier. Some of the environmental issues that the manufacturer considers in selecting suppliers are:

- Does the supplier have established environmental guidelines and practices? It is important to know the record of the supplier in terms of environmental practice and how that aligns with the manufacturer's practice and also industry practices. Specific areas to look at include the nature of materials, process technologies, and energy resources that are used by a supplier in its production process. In fact, is the supplier's production system efficient and can it be reasonably brought to compliance to meet the environmental goals of the manufacturer?

- Does the supplier have the capability for reverse logistics? In other words, when it is time to disassemble and reuse a component, are the products or parts designed with ease of assemble and would they be easily reused or recycled without creating a major environmental burden. Thus, the environmental costs of disassemble and reuse should be lower than the cost of disposal.

- Does the manufacturer adopt a packaging and shipping standards that are consistent with recycling and reuse efforts? And are the materials used easily recycled and biodegradable? Packaging constitutes a major cost of many products today but more and more industries and companies such as LL Bean are developing efficient packaging systems that help to conserve resources such as paper products. It is important that the supplier adheres to a packaging standard that is consistent with sustainable development. For example, many companies are reducing their use of corrugated packages and are also recycling packaging materials.

- The supplier should also have a program for continual improvement in its selection and use of raw materials, technological processes, and packaging. It should be involved in ongoing training activities for its employees and adopt new environmental standards as the laws and guidelines change. Such efforts may help to reduce the use of ozone-depleting substances or the emission of greenhouse gases such as carbon dioxide.

- Suppliers that meet the environmental targets of environmentally conscious manufacturers should be those that have world-class quality performance and are able to minimize rejects, reworks, and scraps. These actions conserve energy resources by not re-running the production line to fix problems and also conserve material and natural resources by not wasting materials through rejects and scraps. Also, being a world-class supplier means that there would be a reduction in emission of gases and disposal of hazardous wastes since these are significantly cut down through efficient production system. In addition, world-class suppliers should have a means of reclaiming hazardous materials that may occur for safe disposal.

- Product design is a critical component in selecting suppliers. Suppliers should design products based on customer needs and must integrate customers in the design stage. However, it is very important to take a holistic outlook of the entire product design phase. This would require developing a life cycle assessment of the product. We have devoted a whole chapter on life cycle assessment and this goes to ensure that the "better" alternative for product design and development is selected after a complete environmental impact assessment. Thus, design for the environment is a necessity.

- Eco-labeling is increasingly getting attention. Many countries, industries and professional associations have adopted labeling schemes. There are two classifications of eco-labeling schemes notably voluntary and mandatory. Popular among them are the German Blue Angel scheme, which became operational in 1977 and the White Swan, which is used in the Nordic countries. In the US, the Green Seal is popular although not endorsed by the government but run by a private organization. These schemes play a role in green consumerism and tend to give the impression of compliance to environmental guidelines. Although they are sometimes misapplied and misused, they could motivate suppliers to meet certain environmental guidelines.

- Distribution and logistics play a key role in selecting suppliers if there is a strong interest in minimizing environmental burden. It would therefore be important to find out how the supplier transports products and how efficient that system of transport and distribution is. The manufacturer may also want to know the measures that are adopted in the transportation of hazardous and toxic materials when they may be involved, the handling of such materials, and the selection and training of workers who handle such materials.

- Finally, the supplier must show sensitivity to environmental practices in each of its activities. The culture for environmental protection must be enshrined in each worker to achieve the goal of environmentally friendly practice. For example, electronic mails may be used instead of snail mail; memos may be sent electronically rather than on printed paper; meetings may be conducted via video conferencing, teleconferencing or through chat rooms rather than on a physical venue; and significant amount of paperwork can be trimmed by streamlining operations. It is therefore, not a simple task to select an environmentally conscious supplier.

### Product Stewardship Practice

Product stewardship requires a change in attitude and manufacturing practices with a focus on conservation of materials and resources. As noted by NPSC, product stewardship practice requires that producers minimize the impacts of their activities on the environment by doing the following:

- Use renewable resources or resources that can be replenished. For example, the use of alternative energy supply such as solar energy or windmill as opposed to the use of fossil fuels. Fossil fuels are limited natural resources and are nonrenewable. Furthermore, the process of excavating and mining them create more environmental burden to the environment.
- Use of biodegradable materials. These are materials that break down into the soil without emitting harmful chemicals or toxic materials to the entire ecosystem. For example, some of the plastic materials are biodegradable and therefore environmentally friendly.
- Use of recycled and/or recyclable materials. Paper products and packaging materials represent a large bulk of recycled and recyclable materials. Many of the paper products that are used now contain a significant percentage of recycled material.

- Use of low or no toxic materials. This is required to avoid the emission of poisonous or toxic gases to the atmosphere and also to avoid the emission of greenhouse gases or ozone-depleting gases.
- Use of sustainable harvesting methods. Land is one of the limited natural resources on earth. There are not enough lands for landfills hence the need to conserve more land. Harvesting activities can also erode the topsoil and deplete the quality of the land. New efforts are made to change the harvesting methods such as cutting down on the use of synthetic fertilizers and encouraging the use of organic bio-stimulants to regenerate the natural condition of the soil. Thus, the use of composting methods is getting increasingly popular.

### Strategies for Product Stewardship

In summary, some of the strategies for environmental management are outlined below:

- Designs for ease of disassembly — Manufacturers are increasingly designing their products so they could be easily disassembled. This has the added advantage of making it easier to recover reusable materials that could be transferred and used in the manufacture of other products. Personal computers represent good example of products that are designed for ease of disassembly. This makes it easier to recover precious materials that could be used in manufacturing new computers.
- Use of modular design— the use of modular design helps prolong the longevity of a product. Hardware systems that are modular designed can be upgraded at the end of their useful life by adding new components that could enhance their continuing performance. So, rather than worrying about the product becoming obsolete, it could be upgraded to continue meeting future challenges. Computer systems are increasingly designed this way.
- Design for dematerialization—this is a process of designing that allows materials to be taken out of the product without affecting the

performance of the product. Products could therefore be restructured or resized without affecting their performance.

- Design for conservation—Emphasis of all design strategy should be on how to conserve materials and energy and minimize waste and any form of pollution.
- Lease options— many products are now offered by manufacturers under lease option rather than outright purchase. The effect of this is that a consumer with short-term need for the product could use the lease option and allow the manufacturer to assume responsibility for the product at the end of the lease. The manufacturer has the necessary support service and logistic network to ensure that the product finds alternative usage at the end of the lease. Also, it puts the disposal responsibility to the manufacturer who has the resources to ensure disassembly, recovery and safe disposal of the product.
- Product take-back—some manufacturers have designed systems that allow recovering the product from the consumer at the end of the product's life. This strategy ensures environmentally compliant disposal and management of the product thereafter.
- Design for recycle—Corporations such as IBM has used 100% recycled plastic in all their personal computer plastic parts. Some others have used 100% recycled materials for their packaging. It is important that parts and products are designed with materials that are recyclable.

## Conclusion

It is clear that for businesses to compete effectively in today's economy, they need to focus attention on the needs and wants of customers and in particular stakeholders. The management of the environment has become an important strategic factor in achieving competitiveness. When environmental quality is not achieved, the cost to both the consumer and the society is significantly increased. The high cost of noncompliance has made many firms rethink their strategies and increasingly, corporations are beginning to realize that it pays to be environmentally conscious. Many have reported significant savings and increased profits as in some of the cases presented in this book. Furthermore, by focusing on environmental protection programs, they have enjoyed the trust and respect of their communities. Firms must accept product stewardship as a major responsibility. We have outlined in this chapter some of the strategies to take to accomplish that goal. We have also noted that environmental compliance is only attainable if all the major participants such as distributors, suppliers, consumers and others participant in the program. It is therefore a systemic problem that needs to be addressed by all that is involved.

## References

Allan, S., "What happened at Love Canal? Alfred, NY: Alfred University. http://cems.alfred.edu/students98/allansm/Onemoretry.html.

Bullard, R.D., Unequal Protection: Environmental Justice & Commentaries of Color, San Francisco, CA: Sierra Club Books, 1994.

"Defining product stewardship," Northwest Product Stewardship Council, http://www.govlink.org/nwpsc/DefiningStewardship.htm, downloaded on September 28, 2001.

Gibbs, L.M., Love Canal: Twenty years later, Harlem Adams Theatre; CSU, Chico: Associated Students Environmental Affairs Council and Environmental Studies Program.

Nersesian, R., "A comparative analysis of Japanese and American production management practices, pp. 37-71, in Management of New Technologies for Global Competitiveness (ed., C.N. Madu), 1993.

Madu, C.N., Managing green technologies for global competitiveness, Westport, CT: Quorum Books, 1996.

# Index

3P, 149
4m, 99
4ms, 120

ABC rule, 117
Acidification, 119, 121
Act, 77
Action Implementation, 131, 133
Agile, 199
Agile manufacturing, 190
Analytic Hierarchy Process (AHP),
    132, 158
Aramis, 48

Benjamin Moore, 2
Bhopal, 9, 89, 91, 230
Blue Angel, 25, 235
Bristol-Myers Squibb, 227
Brundtland, 15, 228
Brundtland Report, 18
Business Charter for Sustainable
    Development, 18

Car of the Year Awards, 10
Carbon emission, 17
CERES, 87
CFCs, 45
Change management, 108
Check, 74
Classification and characterization,
    121
Climate change, 17
Clinique, 48
Compensation, and Liability Act
    (CERCLA), 149
Competitive Benchmarking, 109
Concurrent engineering, 163
Connecticut, 224
Consequence assessment, 219
Conservation Law Foundation, 106
Conspicuous conservation, 3

Corporate image and social
    responsibility, 110
Cradle-to-Grave, 25, 129, 156
Critical Incident Approach, 154
Cross Functional Teams, 182, 184
Customer/Supplier/Vendor
    Relationship, 197

DDT, 45
Dell Computers, 199
Dematerialization, 159
Deming, 4, 74, 76, 95
Design for conservation, 238
Design for dematerialization, 237
Design for disposability, 31
Design for environment, 126
Design for maintainability/
    durability, 126
Designs for pollution
    prevention, 127
Designs for recyclability, 126
Design for recyclability, 31
Design for recycle, 238
Design for remanufacture, 31
Design Strategies, 152
Design team, 152
Designing for environment, 153
Designs for ease of disassembly,
    237
Development, 18, 63
DMADV, 180
DMAIC, 176
Do, 76
Draft International Standard, 29
DuPont, 45

Earth Summit, 15, 63
Ecoefficiency, 19
Ecofactory, 42
Eco-labeling, 25
Ecopoint, 120

Energy star, 150
Environmental Assessment Survey, 154
Environmental Auditing, 8, 63
Environmental Cost, 110
Environmental Defense Fund (EDF), 106
Environmental Labeling, 26, 63
Environmental Liability Laws, 149
Environmental Performance Evaluation, 214
Environmental Performance Indicators, 214
Environmental priority strategy, 120
Environmental Risk Assessment, 210
   Ecological Risk Assessment, 210
   Health Risk Assessment, 210
Environmentally conscious manufacturing, 42
   Zero waste, 42
Estee Lauder, 48
Event tree analysis, 217
Exposure assessment, 218

Failure mode and effect analysis (FMEA), 217
Fault tree analysis, 217
Final Draft International Standard, 29
Fishbone diagram, 99, 120
Flow Manufacturing, 190, 206

General Electric (GE), 183, 223
Geographical Variations, 117
Global warming, 11, 17
Green Accounting, 215
Green market, 150
Green power, 2
Green Quality Function Deployment, 169

Green Seal, 2, 235
Greenhouse Effect, 11

Hazard identification, 217
Hazard indices, 217
Hazards, 210
Hooker Chemical, 231
Horizontal Product Realization, 167
Housatonic River, 223, 227
Hybrid cars, 3, 9

Identify-design-optimize-verify (IDOV), 180
Impact analysis, 30
Improvement analysis, 30
India, 17
Industrial Waste, 24
International Chamber of Commerce (ICC), 19
International Paper, 55
Inventory analysis, 30
Inverse manufacturing, 20, 57
Ishikawa diagram, 158
ISO, 7, 26, 62
ISO 9000, 26
ISO 14000, 26

Japanese EcoMark, 25
Just-In-Time (JIT), 192

Kanban, 204
Khang Hsi, 10
Kodak, 9, 32
Kodak Single-Use Camera, 32
Kyoto, Japan, 17
Kyoto Protocol, 151

Lead Poisoning, 13
Lean Manufacturing, 200, 204
Lean Six Sigma, 185
Lease Options, 238

Life Cycle Assessment (LCA), 25, 115, 130
Life Cycle Cost Assessment, 135
  Appraisal costs, 136
  External failure costs, 135
  Internal failure costs, 135
  Prevention costs, 135
Life Cycle Impact Assessment, 118
Life Cycle Inventory Analysis, 115
LOHAS, 3
Love Canal, 231

Machine, 100
Man, 100
Management of change, 195
Manhattan Project, 231
Mass Customization, 199
Material, 100
McDonald's, 106
Merck & Co, 213
Methods, 102
Modular design, 58
Modular Product Structure, 161
Motorola, 184
Motorola University, 183
Multiple Function Capability, 160

National Association of Attorney Generals [NAAG], 2
National Pollutant Discharge Elimination System (NPDES), 225
National Soft Drink Association (NSDA), 53
Natural Environment, 197
Natural Resources Defense Council, 106
New Economic Order, 16
New England Electric System, 106
Normalization, 121
Nuclear Regulatory Commission (NRC), 213
Nutrification, 119

Occidental Petroleum and Chemical Corporation, 231
Oil Pollution Act (OPA), 149
Organizational change, 196
Organizational culture, 102
Organizational vision and mission, 107

Pacific Gas and Electric Company, 106
Pareto Chart, 156
Plan-Do-Check-Act (PDCA), 95, 79
Planning, 93
Polychlorinated biphenyls (PCBs), 224
Preplanning, 131
Prince William Sound, 89
Product Content Environmental Questionnaire, 155
Product Delivery Process (PDP), 83
Product development, 196
Product stewardship, 23, 236
Product take-back, 238
Pull manufacturing, 204
Pull system, 204

Quality Function Deployment (QFD), 97, 105, 169

Rapid Prototyping (RP), 191
Recyclability, 60
Recycling, 21, 128, 170
Recycling statistics, 52
Reduced Cycle Time, 147
Reduced Regulatory Concerns, 148
Remanufacturing, 57, 100, 168
Resources Conservation, 149
Responsible Manufacturing, 14
Reverse Engineering (RE), 191
Reverse logistics, 23, 43

Reverse supply chain management,
    43
Rio de Janeiro, 15
Risk Assessment, 209, 210
Robin Hood Effect, 16

Schmidheiny, 18
Schulman, 2
SETAC, 117
Six Sigma, 175, 177
Social Responsibility Function, 151
Source-reduction strategies, 44
Stakeholder requirements, 153
Strategic cycle, 93
Strategic environmental
    management, 91
Strategic Information Management
    System, 111
Strategies for Sustainable
    Manufacturing, 20
SUC, 33, 34
Supplier participation, 233
Supply Chain Management, 193
Sustainable Development, 18
Sustainable Manufacturing, 14, 18
SWOT, 107, 196

Table of Impacts, 116
Top Management Commitment,
    106
Total Environmental Quality
    Management (TEQM), 7

Total Productive Maintenance, 205
Total Quality Control (TQC), 205
Total Quality Management, 4
Toyota Production System, 200
Traditional Manufacturing Process,
    196
Type I, 27

U.S. Clean Air Act, 45
Union Carbide, 89
United States Environmental
    Protection Agency, 50
Used Beverage Can (UBC), 53

Value engineering, 160
Vertical Product Realization, 165,
    166
Volatile Organic Compounds, 1
Vulnerability models, 219

Waste-free design, 35
World Business Council for
    Sustainable Development
    (WBCSD), 19
World Class Manufacturing, 198
World Industry Council for the
    Environment (WICE), 19
World Wildlife Fund (WWF), 82

Xerox, 35